The author explores the controversial prediction that an earthquake triggered by planetary tides will destroy Los Angeles in 1982, and the sun-moon alignments which, combined with high winds, could cause repeats of the Blizzard of '78. He reviews some of the specific disasters caused by tides, and explains tidal waves, or *tsunami*, that cause some of the horrors. Here, too, are the military operations, from Caesar's time to our own, that were affected by tides, and some of the other marvels, such as currents, bores and maelstroms.

Altogether, this is a fascinating and factual report of man's effort to understand and anticipate the actions of the waters around the world. Tides can be predicted by computer with precision, but still they often bring disaster, because, as Wylie quotes Professor Paul Godfrey, "Nature always bats last."

Photo by Richard M. Wylie

FRANCIS E. WYLIE took his degree from Indiana University. He spent many years with M.I.T. as PR Director, specializing in interpreting science and engineering news. Earlier in his journalistic career he was the Kentucky correspondent for *Time* and *Life*, until he was tapped to open a news bureau for Time Inc. in Boston. There he covered New England news of all kinds, from the career of Ted Williams to the Brink's Robbery. He now lives in Hingham, Massachusetts.

TIDES and the Pull of the Moon

TIDES

and

the Pull of the Moon

by Francis E. Wylie

The Stephen Greene Press

Brattleboro, Vermont

for John, Kim, and Jeffrey Wylie

PUBLISHED MAY 1979
Second printing September 1979

This book has been produced in the United States of America. It is published by The Stephen Greene Press, Brattleboro, Vermont 05301.

Library of Congress Cataloging in Publication Data
Wylie, Francis E.   1905-
    Tides and the Pull of the Moon.

    Bibliography: p.
    Includes index.
    1. Tides. I. Title.
GC301.2.W94   551.4'708   79-1184
ISBN 0-8289-0347-6

Contents

Preface

READER, if you never have watched the great, golden ball of the harvest moon rise out of the mists; if you have never seen moonlight on midsummer honeysuckle, almost as intoxicating as the scent; if you have never walked under a full moon in January, when the snowdrifts glitter and the shadows are so deep that they seem to be frozen waves; if you have never observed the moon's sparkle on the surf as the tide rolls in under an onshore breeze — in short, if you have never marveled at the magic of the moon and wondered at its power, this book is not for you.

I confess to being a little moonstruck myself. (My great-great-grandmother's maiden name was Moon, which may, in part, account for it.) This was to be expected back in youthful days when moon was rhymed with June and spoon and I could write a poem (happily unpublished) lamenting the discovery that it was not a romantic orb but only a cold, dead clinker — a discovery that only callow disillusion could invest with adequate anguish. In later years I have remained fascinated by the moon, the remarkable phases through which it passes and the infinite variations in our weather through which it reveals itself. At the same time I have watched, at the foot of Steamboat Hill where I live, the coming and going of the tides — the most visible manifestation of lunar power.

Having done my boyhood swimming in the Ohio River, the Wabash, and Bean Blossom Crick, where high water comes from heavy spring rains, I became an adult perhaps more curious about the tides than some who grow up on salt water and take them for granted. Tide tables provide a schedule of ebb and flow but do not explain why the tides vary so much in time and height and at various localities.

In coastal waters, the sovereignty of the moon cannot be ignored. I recall a confident start in a friend's sloop for an all-day race. The skipper was going to outsmart the fleet with a tack close to shore, but the tide was ebbing and the keel ploughed into the mud a few hundred yards from the starting line and in full view of spectators. There we sat, deck canted at 45 degrees, for some four hours until the incoming tide lifted us off. In such a situation, one has cause to ponder the power of the moon.

When the Stephen Greene Press asked if I would be interested in writing a book on the tides I welcomed the opportunity to hunt for answers to questions that have intrigued me, including several about the nature, the behavior, and the influence of the moon not directly related to the tides. *Tides and the Pull of the Moon* was written on the assumption that there are readers who would like to share an equally broad view, on the same level as mine. This level, it will be apparent, does not give access to a complete knowledge of the moon and tides. Complete knowledge is not possible for anyone at this date, and an understanding in depth requires years of study of mathematics, physics, astronomy, and other fields (there are many specialized, scholarly works). I have tried to pull together, for laymen having a general interest, information from science, history, and marine lore about lunar and tidal phenomena and their influence on our life.

Like any other technical field, study of the tides has a language of its own. The *Tide and Current Glossary* issued by the National Ocean Survey defines nearly 500 terms. Some technical words cannot be avoided, but when reasonable I have preferred to use the language common among salt-water people. For example, everyone who has dug clams, sailed a boat, or dipped a toe in the surf speaks of "high tide" and "low tide" though the Glossary states, with chilly sternness, that use of such terms "is discouraged." The approved terms are "high water" and "low water." "Ebb" properly refers only to the direction of the current, though "ebb tide" is certainly in popular use. A fisherman is likely to speak of the tide as "coming" or "going" and of the "top of the tide" or the "bottom." Officially, that period at high or low water when there is no perceptible change in the height is known as "stand of the tide," but I have never heard that term on the waterfront. I have used the term "tide waves" rather than "tidal waves," which can be confused with tsunami.

The British use the term "stream" in referring to what we call a tidal current. To them, "flow" is the combination of the tidal stream and the nontidal current. No doubt there are many other differences in terminology throughout the world.

"Mean range" is a term we cannot avoid and it is used frequently in this book in describing the magnitude of tides in various places. To define it exactly would require a good deal of explanation. For all practical purposes, it is the difference between the average high tide and the average low tide. (This definition is an example of how, in simplifying, I may be guilty of imprecision. I apologize to the experts who, through their books or in person, have helped educate me.)

Some readers may be annoyed that I have not complied with the policy of phasing into the metric system. Aside from the fact that most of us still think in feet, miles, gallons, and pounds, my justification is that the U.S. government measures tides in feet (with fractions in tenths). Foreign tide tables and scientific data are generally in meters, and in most cases I have translated figures from such sources. The velocities of currents are customarily given in knots, one knot being one nautical mile (or 1.151 statute miles) per hour.

In rather extensive notes at the end of each chapter I have listed many individuals and authors to whom I am indebted for assistance. The bibliography will also guide those readers who would like to learn more about the moon and tides.

F. E. W.

Hingham, Massachusetts, 1978

I

Star of Our Life

The moon is not unjustly regarded as the star of
our life. This it is that replenishes the earth; when
she approaches it, she fills all bodies, while, when
she recedes, she empties them. . . . During sleep,
it draws up the accumulated torpor into the head
. . . and relaxes all things by its moistening
spirit.   PLINY THE ELDER

IN 1524 London astrologers pre-
dicted that on February 1 a monstrous tide in the Thames River
would wash away 10,000 houses.

There was panic. At least 20,000 people fled to higher ground by
the middle of January. The prior of St. Bartholomew's not only
erected a refuge on a suburban hill but took along boats and expert
oarsmen, just in case the water came higher than expected.

The public had been assured that the river would rise slowly,
providing a chance to escape, and crowds lined the Thames on
February 1 to watch. Normal tides (they run as high as twenty-four
feet at London) flowed and ebbed. Then the astrologers found that
they had made a slight error. The catastrophe would occur in 1624,
not 1524. (London passed safely through that year also.)

Astrologers, alchemists, and prophets thrived during the early

1500s. This was the period of Nostradamus, Agrippa, and Paracelsus (on whom Dr. Faustus was modeled). Death, in many hideous forms, was never very far away, and the credulous public had good reason to believe the direst forecasts.

The world has changed a great deal in four and a half centuries. Of course, disasters still occur, and we still have prophets and those who believe them. But nowadays, we are better informed and most people live more comfortably than ever before. We have more knowledge of the universe and more certainties about the nature of life (just as many uncertainties, too, but they are different ones).

Few things are more certain than the movements of the moon and the tides they produce, though we must regretfully acknowledge that our knowledge of the certainties of the tides and our technological sophistication together are still powerless to exert more than the slightest influence on the workings of nature. There are still tidal effects that can bring unexpected disaster. Swollen by storms, the tides can flood the land, as London feared. Generated by earthquakes, they can create catastrophe. There is evidence that tides may be involved in the *production* of earthquakes as well. In fact, a prediction has been made that Los Angeles will be destroyed in 1982 by an earthquake triggered by planetary tides. Science was invoked to support the prediction, and, for better or worse, astrologers are said to concur.

Because the sun and moon dominate our sky and the rhythms of nature, man has always revered them and even regarded them as deities. At the very least, they were reliable timepieces. The moon's dependability could be discerned only through a curious pattern of changes through the month, and its heatless light was associated with the dangers of darkness; so it was looked upon as mysterious entity—undoubtedly feminine. Artemis, the chaste huntress, and Hecate, goddess of black magic, were identified with it.

The moon has not always been regarded as feminine. In some primitive cultures, the moon was believed to be the seducer of pubescent girls, bringing on menstruation. The Maori of New Zealand call the menstrual period *mata marama*, meaning moon sickness, and a French euphemism is *le moment de la lune*. There

2

has often been an assumption that it is no mere coincidence that the moon's month and the "monthlies" are equal—a point I shall discuss later.

The crescent moon has often been thought of as a symbol of the crotch. Roman women put silver crescents in their shoes to make sure they would have healthy children. And of course it is still lucky to see the new moon over the left shoulder. (When you see the new moon for the first time, be sure to bow to it three times, or three times three. That superstition may date all the way back to the Egyptian trinity—Osiris, Isis, and Horus.) Strictly speaking, we cannot see the new moon, for it occurs when the moon is between the earth and the sun, which lights only its far side and dazzles our vision. It is usually two or three days later before we see, in the evening sky, the crescent that we refer to as the new moon.

Tradition says more babies are born when the moon is full (statistics, indeed, indicate that this may be true) and to be born at flood tide is lucky. That is also the best time to make butter, Breton peasants thought. And "everyone" knew that milk drawn from a cow during the flood tide will boil over when it is cooked. Cows would burst if fed on clover that was sown during an ebbing tide.

If the flood was the lucky time to be born, the ebb was the proper time to die, according to English lore. Sir John Falstaff died "at the turning o' the tide," and this was widely believed to be the moment when those with chronic or acute disease expired. Samuel Eliot Morison wrote:

> Just as farmers regulated plowing and sowing by the phases of the moon, so sailors and fishermen believed that flood tide meant strength, and ebb tide, weakness. If an old salt lay at death's door, his family and friends watched the tide. If he survived an ebb he would improve with the flood, but he would always die on the ebb.

Writing of his experiences in Civil War hospitals, Walt Whitman said,

'He went out with the tide and the sunset,' was a phrase I heard from a surgeon describing an old sailor's death under peculiar gentle conditions.

I formed the habit and continued it to the end, whenever the ebb or flood tide began the latter part of the day, of punctually visiting those at that time populous wards of suffering men. Somehow (or I thought so) the effect of the hour was palpable. The badly wounded would get some ease and would like to talk or be talked to . . . deaths were always easier; medicines seemed to have better effect when given then.

Among the Haida Indians of the Pacific Coast, a dying man would see a canoe manned by his deceased friends, who would say, "Come with us now, for the tide is about to ebb and we must depart."

As long as 30,000 years ago a Cro-Magnon gouged a series of holes in a reindeer bone which Alexander Marshack, who has made a study of such things, takes to be a record of the passage of the moon through 2¼ lunar months. An almanac? A calendar? A magic implement? No one knows, but perhaps it was all three. The moon was identified with the passage of time, with birth, death, and fertility, with the changing of seasons, with good hunting and times of empty bellies. And the man who could foretell the rhythms of nature must have been regarded as a potent sorcerer.

By the time of the Babylonians, some 5000 years ago, extensive knowledge had been gained of the sun, moon, planets, and stars. The trick of prophesying their movements and human events—part science, part art, and part hocus-pocus—had been formalized. After the Chaldeans conquered Babylon in 540 B.C. astrology dominated religion and culture. The zodiac, that broad band along the ecliptic through which the sun, moon and planets appear to move from one constellation to another, was recognized as a chart of human destiny.

Not the least of the consequences of the conquest of Persia by Alexander the Great was that Chaldean astrology was brought back to Greece and blended with Olympian mythology. The planets themselves were gods and they determined man's fate. Roman religion in turn embodied the astrological gospel and was grafted on the beliefs of northern Europe. If anyone questions that our astrological heritage is substantial he need only recite the days of the week: Sun's Day, Moon's Day, Tiu's Day (for the Norse equivalent of Mars; it is *Mardi* in French), Woden's Day, Thor's Day, Freya's Day, and Saturn's Day.

The heritage is apparent in the great mass of moon lore. The oldest known work on astrology, *Astronomica*, published by the Roman poet Manilius in A.D. 10, has the earliest lucid reference to the Zodiac Man, the Man-of-Signs, the Moon's Man:

> The Ram defends the Head, the Neck the Bull,
> The Arms, bright Twins, are subject to your Rule . . . .
> The parted Legs in moist Aquarius meet,
> And Pisces gives Protection to the Feet.

Zolar, the modern astrologer to whom I am indebted for this excerpt, supposes that Ptolemy, "the greatest astrologer and astronomer of all time," was familiar with Manilius. Ptolemy was born in Alexandria about A.D. 100, and developed from Egyptian, Greek, and Roman thought an earth-centered astronomical concept that dominated science for some fifteen centuries.

In defense of their art, astrologers like to point out that the great early astronomers—Ptolemy, Copernicus, Tycho Brahe, Kepler, and others—practiced astrology. This neither dulls the brilliance of their scientific accomplishments nor proves that astrology is a science. Chemistry owes much to the early work of alchemists, but no chemist today tries to turn lead into gold. Science has learned how to transmute the elements, but not with the Philosopher's Stone, and for other purposes than the creation of precious metal.

Not only does astrology still thrive, but in the past decade it has enjoyed an incredible renaissance. A wonder of the twentieth century is that in a new generation bent on challenging follies and deceits of the past ("Tell it like it is!") a substantial flock has been led into the ludicrous mysteries of astrology and other realms of the occult—with all the paraphernalia of horoscopes, incense,

because it is closest to Earth, but because it controls so many qualities and because its radiation, such as it is, has so many traceable effects
on human life.

Alarmed at the eagerness with which young people have taken
up astrology in recent years, Bart J. Bok, a noted astronomer, led
an unusual movement in 1975—publication of a manifesto,
signed by leading scientists, warning against such gullibility. In
an essay on the subject, Bok wrote:

> The known forces that the planets exert on a
> child at the time of birth are unbelievably small.
> The gravitational forces at birth produced by
> the doctor and nurse and by the furniture in the
> delivery room far outweigh the celestial forces.
> And the stars are so far away from the sun and
> earth that their gravitational, magnetic, and
> other effects are negligible.

Bok didn't mention the moon's gravity, but this also is negligible so far as the individual is concerned. The moon's gravitational
force is only 1/10,000,000th of the earth's; when the satellite is
overhead it does not change the weight of a man by more than the
weight of a drop of sweat.

As we shall see, the moon is responsible for some important
consequences on earth, but to believe it can influence an individual's character and fate requires credulousness of the kind that
has sustained the sale of almanacs for centuries. The first
almanac known to have printed the Moon's Man—the Man of
Signs—was published in 1300. The *Kalender of Shepheards*,
which appeared in Paris in 1493 and London in 1506, had three
pictures of the man, with indications of how various parts of the
body were controlled by the twelve signs of the zodiac. It gave the
medical advice that "a man ought not to make incysyon ne touche
with yren ye membre gouerned of any synge the day that the
mone is in it for fere of to grete effusyon of blode that myght happen, ne in lykewise also when the sonne is in it, for the daunger &
peryll that myght ensue."

Almanacs have continued to be a source of rude medical ad-

8

vice—such as that ulcers, measles, spots on the face, cataracts, epilepsy, and dysentery are lunar diseases. *The Old Farmer's Almanac,* which still publishes the disemboweled man, with zodiacal advice as to the best time for shearing sheep and getting a hairdo, has printed many accounts of medical wonders since it was established in 1792. Cotton Mather, that estimable authority on all things, was the source for this story in the Almanac:

> One Abigail Eliot had an iron struck into her head which drew out part of her brains with it: a silver plate she afterwards wore on her skull where the orifice remain'd as big as an half crown. The brains left in the child's head would swell and swage [bend], according to the tides; her intellectuals were not hurt by this disaster; and she liv'd to be a mother of several children.

The Almanac published an account by Robert Boyle, distinguished seventeenth-century scientist, of an "intelligent person" who broke his head in a fall, so that large pieces of skull were taken out:

> For divers months, that he lay under the chirurgeons hands, he constantly observed, that about full moon, there would be extraordinary prickings and shootings in the wounded parts of his head, as if the meninges were stretched or pressed against the rugged parts of the broken skull. . . . And this gentleman added, that the chirurgeons, (for he had three or four at once) observed from month to month, as well as he, the operation of the full moon upon his head, informing him, that they then manifestly perceived an expansion or intumescence of his brain, which appeared not at all at the new moon, (for that I particularly asked) nor was he then obnoxious to the forementioned pricking pains.

The very word lunatic (from the Latin *lunaticus* for "moon-struck") indicates how ancient is the belief that the moon affects the mind. Folklore says that lunacy is worse and crime, particularly arson, commoner at the time of the full moon; and there is some evidence to back folklore up here. Paracelsus taught that the brain is the moon of the microcosm and thus subject to phases.

After smothering Desdemona, Othello observed:

> It is the very error of the moon;
> She comes more near the earth than she was wont,
> And makes men mad.

Belief in the influence of the moon on the weather and on husbandry has been even more prevalent. The moon affected all fluids, meat as well as human brains. *The Old Farmer's Almanac* advised for January 14, 1794: "Kill your winter pork and beef and it will enlarge with cooking."

In April: "Wheat, sown at this quarter of the moon, is not subject to smutting."

In January 1799: "At this quarter of the moon cut fire-wood, to prevent its snapping."

One of the earliest American almanacs, started by John Tulley in Boston in 1687, was the first to attempt to predict the daily weather, also offering such sayings as "When the small stars are obscured at night, beware of tempests in the offing." An old rule is "A pale moon doth rain." Obviously if clouds are moving in over the stars and moon, the chances of rain are greater than if the celestial bodies are bright. "Red moon doth blow; white moon neither rain nor snow."

"Clear moon, frost soon," is likely to be true in early spring or autumn. As Albert Lee explains in *Weather Wisdom:*

> Farmers, always conscious of crop-hurting frosts, know that when the horns of the new moon are clear, or the face of any phase of the moon is clear, then frost is far more likely than on other nights. This is because on clear nights the cooling of the earth's surface is greatest, and the drop in temperature brings on the condensation on plants we know as frost or dew.

For a long time people speculated about the inhabitants of the moon and there have been some tall stories about them. Sir Wil-

liam Herschel, who discovered Uranus and had other distinguished accomplishments, studied the mountains on the moon.
The Old Farmer's Almanac reported "from a London paper" that
"Mr. Herschel is now said, by the aid of his powerful glasses, to
have reduced to a certainty, the opinion that the moon is inhabited. . . . Within these few days he has distinguished a large edifice,
apparently of greater magnitude than St. Paul's; and he is confident of shortly being able to give an account of the inhabitants."
Herschel's son, Sir John Herschel, also an eminent astronomer,
went to the Cape of Good Hope to study the southern sky; and in
1835 the New York *Sun* published a story—a complete
hoax—that he had observed "winged men" on the moon.

H. G. Wells imagined moon-dwellers as ant-like in anatomy
and social organization, living in an enormous cavern and bringing their "mooncalves" (huge, slug-shaped creatures) to the surface to browse on plants that sprang up spectacularly at the end
of the lunar night. In *The First Men in the Moon* (1901) his
protagonists of the title arrived in a craft propelled by gravity. A
metal alloy which insulated against gravitational force on earth
permitted the craft to be pulled to the moon and it managed to
return the same way, with a window blind sort of contrivance
which could shut off or turn on the alloy's effect.

Roger W. Babson, an engineer who became rich in business
and by advising other people how to get rich, took very seriously
the idea that a substance could be found to neutralize gravity. In
1948, the year he founded the Gravity Research Foundation in
New Hampshire, he predicted that before the year 2000 "There
probably will be found some product which insulates against
gravity. This would make it possible for an individual to stand on
something three feet square and a few inches thick and eliminate
his weight so that he could fly personally, with or without the aid
of a motor." Such an insulator would be a source of unlimited
power and would be very useful in shoes or stairway treads, he
thought.

No antigravity device has been found for going to the moon.
Jules Verne, in 1866, was more of a realist than H. G. Wells. His
voyagers in *From the Earth to the Moon* and *Round the Moon*

were propelled by 400,000 pounds of explosive from a 900-foot cannon sunk in the ground in Florida. This feat was theoretically feasible—at least the location of a launch and the calculated escape velocity were correct. Rounding the moon, the explorers correctly decided that it was not habitable.

Among the seven ways that Cyrano de Bergerac (as described by Edmond Rostand in 1897) conceived of going to the moon was this one:

> I might construct a rocket, in the form
> Of a huge locust, driven by impulses
> Of villainous saltpetre from the rear,
> Upwards, by leaps and bounds.

But Cyrano chose a more poetic way to make the trip:

> The ocean! . . .
> What hour its rising tide seeks the full moon,
> I laid me on the strand, fresh from the spray,
> My head fronting the moonbeams, since the hair
> Retains moisture—and so I slowly rose
> As upon angels' wings, effortlessly
> Upward—then suddenly I felt a shock!

When men actually went to the moon for the first time, the rocket method was the choice, with "villainous" liquid hydrogen and oxygen as the fuel. People in the twentieth century have become accustomed to technological accomplishment, but this was an epic feat that will be remembered through all the centuries to come. For those of us who were there, it was dazzlingly memorable: to arrive before dawn at the Cape Kennedy launch site and see the floodlit 363-foot Apollo 11, its flanks bathed in vapor; to watch, knowing that three men were waiting in the tip to be hurled into space; to witness the great burst of fire at the base and see the rocket slowly lift upward, then hurtle into the blue tropical sky.

By the time of the first manned expedition, the moon had been carefully studied by spacecraft in orbit, but no one was certain just what its surface would be like. One eminent scientist had speculated that it might be covered with dust a mile or so deep and that men landing there would sink below the surface and never be seen again. There was dust, but the surface was more like a damp beach after the retreat of the tide.

Immediately after landing at 9:56 P.M. (Houston time) on July 7, 1969, Neil Armstrong reported to earth:

> It's gray and it's very white chalky gray, as you look into the zero phase line, and it's considerably darker gray, more like the ashen gray as you look up ninety degrees to the sun. The— some of the surface rocks in close here, that have been fractured or disturbed by the rocket engine plume, are coated with this light gray on the outside, but when they've been broken they display a dark, very dark gray interior and it looks like it could be country basalt.

So the moon wasn't made of green cheese. In fact, our satellite failed to fulfill most of the fantasies of the centuries. Yet it is certainly different from the earth (who would have supposed that the soil, in part, is made up of glass beads?) and scientists are still studying the geological samples brought back and have many puzzles yet to solve. In any case, we are mated to the moon and shall be for a long, long time. The hard facts about the moon that are known, and those yet to be learned, are far more exciting than the squashy suppositions of astrology and superstition.

NOTES

The quotation from Pliny the Elder comes from his *Natural History,* Book II, probably published in A.D. 77.

The story of the high tide predicted for London in 1524 is told in *Memoirs of Extraordinary Popular Delusions and the Madness of Crowds,* a remarkable work by Charles Mackay (National Illustrated Library, London, 1852, and L.C. Page & Company, Boston, 1932).

The Golden Bough by Sir James G. Frazier (Macmillan, New York, 1922) remains the great standard reference for folklore. I also drew upon *A Treasury of American Superstitions* by Claudia de Lys (Philosophical Library, New York, 1948) and *Weather Wisdom* by Albert Lee (Doubleday, Garden City, 1976). Samuel Eliot Morison is quoted from his delightful *Spring Tides* (Houghton Mifflin, Boston, 1965); and he had drawn upon *Walt Whitman's Civil War,* edited by Walter Lowenfels (Alfred A. Knopf, New York, 1961).

In my attempt (doubtless biased) to learn something about astrology, I read — or read at — a number of books and found the most useful to be *Astrology* by Louis MacNeice (Doubleday, Garden City, 1964) and *The History of Astrology* by Zolar (Arco Publishing, New York, 1972). I have quoted from *Astrology for the Aquarian Age* by Alexandra Mark (who kindly answered some questions) (Essandess Special Editions, Simon & Schuster, New York, 1970); and *Astrology* by Hans Holzer (Henry Regnery, Chicago, 1975).

For the skeptical point of view, see *Objections to Astrology*, especially the articles by Bart J. Bok and Lawrence E. Jerome (Prometheus Books, Buffalo, 1975) and *Occult America* by John Godwin (Doubleday, Garden City, 1972).

Almanac information and quotations came chiefly from *The Old Farmer and His Almanack* by George Lyman Kittredge (William Ware and Co., Boston, 1904); *America and Her Almanacs* by Robb Sagendorph (Yankee, Inc., and Little, Brown, Boston, 1970); and *The Old Farmer's Almanac Sampler* (Ives Washburn, New York, 1957) edited by Robb Sagendorph, who resuscitated the Old Farmer.

For *The First Men in the Moon*, I had handy *Seven Famous Novels by H.G. Wells* (Alfred A. Knopf, New York, 1934); and for *From the Earth to the Moon* and *Round the Moon*, I had *The Omnibus Jules Verne* (J.B. Lippincott, Garden City, undated). The excerpt from *Cyrano de Bergerac* is from Brian Hooker's translation (Modern Library, New York, 1923).

Miss Elizabeth Di Bartolomeis, Newton Collection librarian at Babson College, helped refresh my memory of interviews with Roger W. Babson. *First on the Moon, A Voyage With Neil Armstrong, Michael Collins and Edwin E. Aldrin, Jr.* (Little, Brown, Boston, 1970) did likewise on recollections of Apollo 11 and provided the Armstrong quotation.

II

Boiling of the Waters

> Beda the priest says the tides follow the moon. . . . When the moon is opposite to the sun, the sun heats the ocean greatly, and as nothing impedes the warmth, the ocean boils and the sea flood is more impetuous than before.   ICELANDIC RIMBEGLA

GRAVITATION is the most pervasive circumstance of existence. After the near-weightlessness of the womb, a baby becomes acclimated to the earth's pull. As he grows older, he learns from experience that he falls down, not up. There is no occasion for him to wonder, Why?

Since all men have grown up in the field of gravity and have taken it for granted, it is not surprising that in the early groping to understand nature we did not recognize clues to the gravitational framework of the universe and to the cause of tides. To have supposed that the moon attracted water would have seemed more fantastic than the theory, cited by Aristotle, that "the sea is a kind of sweat exuded by the earth when the sun heats it, and this explains its saltiness."

Aristotle spent the last months of his life on Euboea, an island separated from Mainland Greece by the Euripe, a narrow channel through which there is a notoriously dangerous tidal current of nine knots. A legend grew, and persisted for centuries, that Aristotle flung himself into the strait, committing suicide because he could not find the cause of the tides.

Tides in most of the Mediterranean are so slight that scientists and sailors of ancient times were little aware of them, but the tidal differential between the Aegean Sea and the Mediterranean Basin is great enough to create the furious current in the Euripe. Aristotle knew, of course, that rivers flow from a higher level to a lower level; he regarded the Mediterranean as a river which flowed into a great, mysterious ocean beyond continental limits. But he was also aware of the backward and forward movement of waters. In *Meteorologica* he wrote:

> It is true that in straits, where the land on either side contracts an open sea into a small space, the sea appears to flow. But this is because it is swinging to and fro. In the open sea this motion is not observed, but where the land narrows and contracts the sea the motion that was imperceptible in the open necessarily strikes the attention.

So Aristotle did not prepare his star pupil, Alexander the Great, for the tides of the Indian Ocean, which can be devastating when driven by monsoons. As he marched along the Indus River in the summer of 325 B.C. during his conquest of India, Alexander decided to take his fastest boats, thirty-oar galleys and lighter galleys, onto the sea. His biographer, Arrian, wrote:

> Near the mouth of the river, where it attains its greatest breadth of some 25 miles, they encountered a hard wind blowing in from the open sea, and the water was so rough that it was hardly possible to lift the oars clear of the crests and they were compelled to run for shelter into a small creek, under their pilots' direction. Lying in the creek, they were caught by the ebb tide and left high and dry. The tide is, of course, a regular feature of the ocean, but the ebb came as a surprise to Alexander's men, who had had

no previous experience of it, and it was an even
greater surprise when, in due time, the water
returned on the flood and the vessels were re-
floated. Some of them had settled on soft mud;
these were floated again without sustaining any
damage and were able to continue their voyage;
but others, which had been caught on a rocky
bottom and were not evenly supported when the
tide left them, did not fare so well: as the flood
came in with a rush, they either fell foul of one
another or bumped themselves to pieces on the
hard bottom.

Pytheas of Marseille, a navigator believed to have been Alexander's contemporary, was the first Greek to gain extensive
knowledge of the tides and relate them to the movement of the
moon. He voyaged to the British Isles and beyond, and reported
seeing tides 120 feet high. Because his own writings have not survived, it is possible that exaggeration may have come through
second-hand versions rather than from an explorer's tall story.

Posidonius, Stoic philosopher of the first century B.C., made
observations of the tides in the Atlantic Ocean at Cadiz and described their variation—twice a day, monthly, and through the
year. Sailors of that time and place must have had such information, but Posidonius recorded it; and although his writings have
disappeared they were a useful source for Strabo, the great geog-
rapher, who was born a few years after his death.

By that time the relationship of the moon and tides was generally accepted. Pliny the Elder, Roman of the first century A.D.,
discovered the lunitidal interval—the delay between passage of
the moon and flood tide. But the means by which the moon caused
tides was still a mystery. Strabo cited the opinion of Seleucus of
Babylon that the moon compressed the atmosphere and that this
produced tides by pressure on the sea. Even fifteen centuries
later, René Descartes was committed to this notion.

There were many weird explanations. In the eighth century,
Paul the Deacon spread the theory that a huge whirlpool sucked
the ocean in to cause low tide and then spewed the waters out for
the flood. In the tenth century, the Arabic writer El-Mas 'údí told

of the belief that the moon heated up the water and caused it to expand. Zakariyya ibn Muhammad ibn Mahmud al Qazvini, a thirteenth-century Arabic writer, stated in *Wonders of Creation*:

> As for the flow of certain seas at the time of the rising of the moon, it is supposed that at the bottom of such seas there are solid rocks and hard stones, and that when the moon rises over the surface of such a sea, its penetrating rays reach these rocks and stones which are at the bottom, and are reflected back thence; and the waters are heated and rarefied and seek an ampler space and roll in waves towards the seashore . . . and so it continues as long as the moon shines in mid-heaven. But when she begins to decline, the boiling of the waters ceases.

The brilliant thinkers of the Renaissance worried about the problem of the tides while struggling with the great, fundamental mystery of how the whole universe worked. Leonardo da Vinci jotted in his notebook: "As man has within him a pool of blood wherein the lungs as he breathes expand and contract, so the body of the earth has its ocean, which also rises and falls every six hours with the breathing of the world." This must have been poetic fancy, for elsewhere he recognized that the sun and moon had a power over the tides and reasoned (erroneously) that the shallow water of swamps should offer less resistance to this power than the deep ocean.

Before the tides could be understood, concepts of orbiting worlds and of gravity had to be developed. As far back as the third century B.C. the Greek astronomer Aristarchus had proposed the wild and blasphemous idea "that the heavens remained immovable, and that the earth moved through an oblique circle, at the same time turning on its axis." Those were the words used three centuries later by Plutarch in his essay "On the Face of the Round Moon." And why shouldn't the moon, also revolving, tumble on our heads? Plutarch went on:

> The moon has, for an help to preserve her from
> falling, her motion and the impetuosity of her

> revolution; as stones, pebbles, and other
> weights, put into slings, are kept from dropping
> out, whilst they are swung round, by the swift-
> ness of their motion.

About fourteen centuries passed before Copernicus, reading this, was encouraged to work out a solution for the great riddle of the solar system's clockworks — how the moon revolves around the earth and how the earth and other planets revolve around the sun.

In the early seventeenth century, Galileo Galilei in Italy and Johannes Kepler, a German astronomer who became an associate of Tycho Brahe at Prague, were ardent advocates of the Copernican system — Galileo at the peril of being burned at the stake by the Inquisition. Through experiments, mathematical analysis, and telescopic observation, they extended the understanding of the universe. Both believed that the sea lagged in trying to keep up with the rotation of the earth.

Galileo, in fact, regarded this movement of the sea as a proof that the earth rotated. He had observed that on the barges carrying fresh water to Venice a sudden movement would cause the water to slop back and forth. He thought the tides were caused by such a movement in the ocean.

Kepler decided the tides were caused by an attraction of the moon and that the sea was not pulled clear away from the earth because the earth had a greater attraction. Galileo scoffed at this idea as that of a medieval astrologer. Indeed, Kepler earned a living by drawing horoscopes, as did many other earnest astronomers in those times, and he was something of a mystic. He later faltered in advancing his gravitational theory.

The idea of gravity had been around for a long time, though no scientist had pinned down the vague concept. The power of the lodestone was known and references to gravitation were made in terms of magnetism. William Gilbert, court physician under Queen Elizabeth, discovered in experiments with a compass that the earth itself was a magnet and concluded that magnetism (the earth's "soul") caused a dropped stone to fall and was involved in the motions of the sun and planets.

In 1650, Bernhardus Varenius struggled with the problem of the tides in his *Geography* thus:

> Some have thought the Earth and Sea to be a living Creature, which by its Respiration, causeth this ebbing and flowing. Others imagined that it proceeds, and is provoked, from a great Whirlpool near Norway, which, for Six Hours, absorbs the Water, and afterwards, disgorges it in the same space of Time. Scaliger, and others, supposed that it is caused by the opposite Shores, especially of America, whereby the general Motion of the Sea is obstructed and reverberated. But most Philosophers, who have observed the Harmony that these Tides have with the Moon, have given their Opinion, that they are entirely owing to the Influence of that Luminary. But the Question is, what is this Influence? To which they only answer, that it is an occult Quality, or Sympathy, whereby the Moon attracts all moist Bodies. But these are only Words, and they signify no more than that the Moon does it by some means or other, but they do not know how: Which is the Thing we want.

When these words appeared, Isaac Newton was only eight years old (he was born in 1642, the same year that Galileo died). He was twenty-two years old and a fellow at Trinity College when the plague broke out in Cambridge and he fled to a farm in his native Lincolnshire. During two years there he finally explicated the gravitational structure of the solar system and solved the riddle of the tides. Whether or not an apple, as first related by Voltaire, was involved, Newton made an amazing mental leap to grasp the fact that the moon was no different from a small object falling to the surface of the earth. Louis Trenchard More wrote:

> As in a vision, he saw that if the mysterious pull of the earth can act through space as far as the top of a tree, of a mountain, and even to a bird soaring high in the air, or to the clouds, so it might even reach so far as the moon. If such were the case, then the moon would be like a

> stone thrown horizontally, always falling
> towards the earth, but never reaching the
> ground, because its swift motion carried it far
> beyond the horizon. Always falling towards the
> earth and always passing beyond it, the moon
> would follow its elliptical path if these two mo-
> tions were equally balanced.

Although Newton had the insight, his development of the mathematical proof went slowly and twenty years passed before he published his findings. *Philosophiae Naturalis Principia Mathematica,* which appeared in 1687, was doubtless the most magnificent treatise in the history of science — in its breadth, in its precise detail, and in the simplicity with which it explained puzzling phenomena.

So far as we are concerned, the important revelation of *Principia* was the Law of Universal Gravity: that two bodies are attracted to each other in proportion to the product of their masses and in inverse proportion to the square of the distance between their centers. This law explained how the tides are caused by the gravitational attraction of the moon and the sun. And the tides, which people at that time were beginning to observe closely, represented a daily demonstration of the correctness of Newton's theory and calculations. But Newton still could not fathom the *nature* of gravity. He wrote:

> Hitherto we have explained the phenomena of
> the heavens and of our sea by the power of
> gravity, but we have not yet assigned the cause
> of this power. . . . I have not been able to
> discover the cause of those properties of gravity
> from phenomena, and I frame no hypothe-
> sis. . . . To us it is enough that gravity does
> really exist and act according to the laws which
> we have explained, and abundantly serves to ac-
> count for all the motions of the celestial bodies
> and of our sea.

We still do not know what gravity is.

Everyone is aware that gravity is the force that causes a bowling ball to drop on a toe, the moon to pull at the ocean, and a comet

be a source of gravitational waves, and if these waves can be detected they will help not only in understanding black holes but in probing to the very edge of the universe.

According to Einstein, an oscillating mass should send gravitational waves through space. Kip Thorne puts it this way:

> A gravitational wave is a ripple in the curvature of spacetime that is ejected from the black hole in its birth throes and then propagates toward the earth with the speed of light. . . .What this ripple of curvature does is to jiggle neighboring inertial reference frames relative to each other. And since matter initially at rest likes to remain at rest relative to its inertial frame, the wave also jiggles adjacent pieces of matter relative to each other.

What one needs to detect such a wave is a mechanical spider to sense the twitch in the cosmic web.

Nearly a decade ago Joseph Weber of the University of Maryland announced that he had detected gravitational waves for the first time. As an antenna, he used a 1½-ton bar of aluminum—a big chunk of metal larger than the oil tank in your basement. The bar was fitted with strain gauges to register any jiggling of its atoms.

Other scientists were skeptical about Weber's report, for gravitational waves would be extremely faint and there are all kinds of terrestrial disturbances which could jiggle the bar. So a twin antenna was set up at Argonne National Laboratory near Chicago, synchronized with the one at College Park, Maryland—the theory being that a passing truck which caused a jiggle at one place would certainly not cause confusing noise at the other. At last reports, scientists were still arguing with Weber about the mathematical analysis of his data.

Weber, Vladimir Braginsky of Moscow University, and David Douglass of the University of Rochester more recently have been experimenting with crystals which, they hope, might be 10,000,000 times more sensitive than an aluminum antenna, especially if they are large. A 200-pound sapphire would be just the thing. It could be manufactured, but it would indeed be a precious jewel.

Other schemes are being developed. Rainer Weiss of the Massachusetts Institute of Technology conceived the idea of using the technique of an interferometer, a device invented nearly a century ago by A. A. Michelson to test whether the speed of light is changed by the movement of the earth. (No change was found, and the experiments helped lay the foundation for Einstein relativity. Michelson won a Nobel Prize.) A beam of light is split and then brought together after being reflected by mirrors, showing by overlapping light waves whether the two paths are identical.

Weiss wants to use four, or perhaps eight, mirrors among which a split laser beam would bounce several hundred times before being recombined and revealing by its wave pattern whether the mirrors had been jiggled by a gravitational wave. To avoid the inevitable vibrations on earth, the mirrors would be suspended several miles apart, way out in space — quite a trick but feasible in satellite technology.

"Such an apparatus would be perhaps a billion times more sensitive than Weber's," Weiss says. "The instrumentation for detecting the vibrations would have to be enormously sensitive. The experiment might cost a hundred million dollars. It could not be done before some time in the 1980s and, although NASA has considered it, there are no definite plans."

Nevertheless, Weiss has built a small prototype of the apparatus for testing the method — small, though it almost fills a laboratory room. He does not have the slightest expectation of detecting gravitational jiggles yet. Ronald Drever in Glasgow and H. Billing in Munich have similar projects.

One reason scientists were skeptical about Weber's experiments was that he recorded gravitational bursts several times a day. And it's not that often that a star explodes or a black hole is formed. Relatively frequent events might be expected in the Virgo cluster, which has about a thousand galaxies. But even then, "You might get one a month if you were lucky," says Weiss.

There has been speculation that gravity has been getting weaker since the creation of the universe — that the sacrosanct gravitational constant is not a constant after all. If this is true, some conclusions about cosmic evolution would have to be changed. The sun may be only half as old as it has been thought to be, Robert H. Dicke of Princeton University has pointed out. The earth may be expanding—which could be a factor in continental drift, climate change, and sea level variation. The moon may be expanding too; and if this proves to be the case, it could explain slight discrepancies in the moon's movement and make its history a little more comprehensible.

Historical records of eclipses through the centuries, revealing changes in the eclipse timetable, could help in determining whether gravity is getting weaker—but for one obstacle. To calculate whether inconsistencies in the eclipse schedule are due to gravitational change, one first needs to know to what extent they are due to the tides, which cause changes in the movements of the earth and moon. A physicist can make an estimate of an "energy budget," making an allowance for all the energy that goes into the movement of the earth and moon and into the tides they create, but not enough is yet known about the effect of tides to make the budget sufficiently precise.

Tidal effects, as we shall see later, have a bearing on the future of the earth, and gravitational change would have implications in the comprehension of the whole universe. So the small excursion I have made into gravity, on the way to discussing ocean tides, seems justified. And it would be fun to know if, a few centuries hence, learned people will consider the concept of the black hole—a sort of gravitational whirlpool in space—as fanciful as the idea of a whirlpool that caused tides by sucking in the ocean.

NOTES

The quotation at the head of the chapter was taken (and condensed) from *The Tides* by George Howard Darwin (Houghton Mifflin, Boston, 1899), a basic book that anyone who studies the subject should read. Darwin credits a Mr. Magnusson for finding it in Icelandic literature. The quotation from *Wonders of Creation* is also found in Darwin.

By far the best history of knowledge of the tides is *Scientists and the Sea* by Margaret Deacon (Academic Press, London, 1971) and I have drawn on it for this chapter and others. The account of Alexander's encounter with the tides is quoted from *The Life of Alexander the Great* by Arrian, translated by Aubrey de Selincourt (Penguin Books, Baltimore, 1958). A readily available edition of Aristotle's *Meteorologica*, Book II, is that published by the Harvard University Press in 1952, the translation by H. D. P. Lee. In addition to the quotation from Leonardo there are several references to tides in *The Notebooks of Leonardo da Vinci*, translated by Edward McCurdy, Vol. I (George Braziller, New York, 1958).

The quotation from Plutarch's "On the Face of the Round Moon" was found in *The Origins of Scientific Thought* by Giorgio de Santillana (New American Library, New York, 1961). Everyone interested in the history of science should read *The Watershed*, a biography of Johannes Kepler by Arthur Koestler, taken by Doubleday, Garden City, 1960, from *The Sleepwalkers* (Macmillan, New York, 1959).

The quotation from Bernhardus Varenius appeared in *The Tide* by H. A. Marmer (D. Appleton and Co., New York, 1926), which, though old, is perhaps the best book on the tides ever written for the layman. Marmer was assistant chief of the Division of Tides and Currents, U.S. Coast and Geodetic Survey. I have quoted from *Isaac Newton* by Louis Trenchard More (Dover, New York, 1962). *Isaac Newton, A Memorial Volume*, edited for the Mathematical Association by W. J. Greenstreet (G. Bell and Sons, Ltd., London, 1927) was another source. The Newton quotation is from *Newton's Philosophy of Nature, Selections from His Writings* (Hafner Publishing, New York, 1953), useful to those who don't care to tackle *Philosophiae Naturalis Principia Mathematica* (1687).

Gravity by George Gamow in the Science Studies Series — (Doubleday, Garden City, 1962) — is an excellent introduction to the subject (if you have a fairly good background in mathematics). The best recent review I found on gravity research, in lay language, was an article, "Probing the Universe: Big Bang, Black Holes, and Gravitational Waves," by Kip S. Thorne, which appeared in CalTech's magazine, *Engineering & Science* (October–November 1977, Vol. XLI, No. 1) and I have quoted from it. *The Key to the Universe* by Nigel Calder (Viking Press, New York, 1977) was helpful. Robert H. Dicke's *Gravitation and the Universe* (American Philosophical Society, Philadelphia, 1970) is informative and it includes chapters by Joseph Weber; all of this material is quite technical.

I owe special thanks to Dr. Rainer Weiss, professor of physics at the Massachusetts Institute of Technology, for stimulating tutelage.

III

Mechanics of the Tides

Compelled by fates that likewise rule the sea,
I roll out month-long periods of time
In sure-returning cycles. As the light
Of glorious beauty slowly leaves my face,
So does the ocean, flowing from the shore,
Lose its increase of waters in the deep.
RIDDLE BY ALDHELM FOR KING ALDFRITH

HAVING MADE a bow to Einstein, we can return to Newtonian physics for an explanation of the tides. Isaac Newton did not have the last word, of course. He was concerned with fundamental principles and calculated the force of the sun and moon on a theoretical sort of earth.

That was a long way from describing the motion of the seas as they actually exist and from explaining the great differences in tides in various parts of the world. One must make allowances for what is obvious to a child: water runs. It runs downhill, seeking the lowest level. It sloshes and splashes and moves in the form of waves when it is disturbed. And tidal forces make it run uphill.

This kind of behavior of water was taken into account by Pierre

Simon Laplace, who has been called the Newton of France. A farmboy who became a marquis through his distinguished mathematical studies, he published the five-volume *Mécanique céleste*, beginning in 1799, and provided a new view of the universe. He was able to explain how the tides occur as global waves, successfully analyzing the complex movement of a rotating fluid under gravitational influences.

Since Laplace's time, scientists have learned a great deal more about the tides, though knowledge of them is still not complete.

To comprehend the tides, one must first know the relationship of the earth to the sun and moon. Perhaps today everyone, having watched space vehicles orbit on television, has a "feeling" for the way in which celestial bodies revolve. Today's children are more familiar with the orbiting of a spacecraft around the moon than they are with the stone Plutarch whirled in a sling, even though practically every teacher since Plutarch has demonstrated gravity the same way. They can easily understand that gravitation keeps the earth on a leash as it revolves around the sun and keeps the moon tethered as it revolves around the earth.

It is less easy for us to understand that actually the moon and the earth revolve around each other. Or, to be more exact, they revolve around a common center of gravity which, because the earth is so much bigger, is about a thousand miles below the surface of the earth — some 3,000 miles from its center. This is an important factor in the tides and should be remembered, but it is simpler to speak of the moon revolving around the earth once a month — or, to be more precise, an average of every 29.53 days, from new moon to new moon. This period is known as the synodical month.

Another point to keep in mind is that while the gravitational attraction between two bodies varies inversely with the square of the distance between their centers, tidal forces vary inversely with the *cube* of the distance between them. The difference is that tidal forces act on the surface of the whole globe. To demonstrate this requires mathematics, and we would do well just to accept it on faith.

The rule governing tidal forces makes it clear why the sun, which is 27,000,000 times bigger than the moon, has less than half the moon's power over tides on the earth. The sun is 93,000,000 miles away and that number cubed is enormously greater than the cube of the moon's distance from the earth — an average distance of 238,855 miles. The sun's influence on the tides is only about forty-six percent of the moon's. This being the case, we are justified in first giving our attention to the moon. One might suppose that the tide would be highest when the moon is directly overhead. Such is not the case. The vertical pull is very slight — only one ten-millionth of the weight of water. If this were the only force, we would have no perceptible tides.

But in addition to this attractive force, there are horizontal forces, known as tractive forces, which are exerted on waters not directly under the moon and pull them toward one point. It is easier to pull a boat across the water than to lift it out vertically. With similiar mechanics, the moon can move prodigious amounts of ocean water.

This is only the beginning of the story. At the same time that the moon is creating a high tide on one side of the earth, a high tide is also occurring on the opposite side — contrary to what common sense tells us should happen. Why should this be?

We must now recall the fact that the earth and moon revolve about a common center of gravity. The old-fashioned chain shot — a naval weapon consisting of two cannonballs chained together and fired at enemy ships to destroy the rigging — must have revolved somewhat in this fashion as it hurtled through the air. The centrifugal force, as the linked earth-moon revolves, tends to hurl the oceans into space — a calamity prevented by the earth's gravitation. On the side of the earth facing toward the moon, and closer to it, the moon's pull is greater than the outward-hurling centrifugal force, and a high tide occurs as we would expect. On the opposite side, farther away from the moon, the moon's pull on the sea is less than the centrifugal force. Furthermore, the moon is pulling all the particles in the earth away from the waters there. So high water comes at about the same moment on opposite sides of the earth. Half-way around the globe, the seas are being pulled away toward those high-water points, and low tide, or low water, occurs.

But the earth is also rotating, and as it rotates various points pass through the two tidal bulges and the two tidal depressions. The earth rotates once in twenty-four hours. Since at the same time the moon is revolving about the earth in the same direction, it takes twenty-four hours and fifty minutes (on the average) for any point to catch up and come beneath the moon again. If there is a high tide at New York at noon today, there will be one at about 12:50 P.M. tomorrow. And half-way between, there will be another at about twenty-five minutes after midnight. These are twice-daily, or semidiurnal, tides.

The two tidal bulges can be regarded as waves. Because of the earth's rotation, the moon appears to travel westward at a speed of 1,000 miles an hour, so one would expect a high-tide wave to travel through the ocean at 1,000 miles an hour. Under the laws of physics, such a speed would be possible if the earth were covered with oceans fourteen miles deep. But the average depth of all the oceans is only two miles. And only in the seas surrounding Antarctica, narrowed in the Drake Passage below South America to 500 miles, is there no obstruction to east-to-west movement. Elsewhere, the tide waves are blocked by continents and islands, slowed down in shallow waters and diverted by all manner of capes and peninsulas before they can reach bays, estuaries, and harbors. The waves are reflected from land and react with the waters over which they return. The result of all this is that tides arrive at different parts of the world in a very complicated pattern of times and heights. The speed of a tide wave is not more than 700 or 800 miles an hour. So the tides get out of step with the moon.

There are other reasons for being out of step, and at this point it would be well to bring the sun back into the picture. Although the sun exerts less than half the moon's gravitational force, its influence is substantial. The two bodies are on different schedules and, depending on their positions in relation to the earth, they may pull in the same direction or in different directions, augmenting or diminishing the height of the tide and changing its schedule.

Twice a month the sun and moon are lined up with the earth in what is known as *syzygy* (an awkward-looking Greek word that is easier to pronounce than it looks — "sizz-a-jy"). This occurs when they are *in conjunction*, on the same side of the earth — at new

At syzygy, when moon, earth and sun are lined up, the sun reinforces the gravitational pull of the moon and spring tides occur. This drawing shows a full moon, but the effect is identical at new moon, when the sun and moon are on the same side of the earth. The arrows on the earth show the "tractive" force which is chiefly responsible for the tides, the length of the arrows indicating where this horizontal force tends to be greatest. *(Drawing by Richard L. Murphy, Jr.)*

moon when the sun is shining on the other side of the moon and it is therefore dark to us; and it occurs when they are *in opposition*—on opposite sides of the earth, at full moon, when the sun's light is reflected by the moon at its maximum. (If they were in a precise direct line we would have an eclipse of the moon and an eclipse of the sun once a month, but usually they are not lined up so exactly.)

When the sun and moon are in conjunction, they reinforce each other's tidal pull—both pulling in the same direction to make the tides higher. The same thing is true when they are in opposition. They are pulling in opposition but, for the reason explained a few paragraphs earlier, they produce tides on opposite sides of the earth that are higher than average. These high tides at syzygy are called *spring tides,* the term coming from the ancient verb *springen* (to leap up) and having nothing to do with the season of spring.

Twice a month the sun and moon are at right angles to each other—*in quadrature.* This is at the time of the first and last quarters, when only half the face of the moon is showing. At these times, the sun's gravitational force is subtracted from that of the moon and we have *neap tides,* which are tides that are lower than average. Tidal force at that time may be only a third of the force at spring tides. The difference is not trivial. At Boston, for example, where the mean range (the difference between mean high water and mean low water) is 9.5 feet, a spring high water can be greater than twelve feet. The lowest neap high water during the current year is eight feet.

Spring tides also bring lower low waters. At Boston in the current year low tide during the springs falls two feet or more below mean low water fifteen times, exposing tidal flats not often seen. During the neaps, on the other hand, low tide is frequently a foot or more above mean low water.

As the moon waxes and wanes through the month there are many variations in heights and times of tides. The basic rhythm, repeated monthly a dozen times a year, is inexorable, but it varies in response to a number of factors.

One important factor is the distance of the sun and moon from the earth. Since the moon's orbit around the earth and the earth's orbit around the sun are elliptical, the distances between earth and moon and earth and sun change. At its closest—at perigee—the moon is at an average of only 221,463 miles away. At apogee—when it is farthest—the distance is 252,710 miles. In the course of a month its tidal force varies about twenty percent from that at the mean distance.

When the sun's distance is the least—at perihelion—the tidal effect is greater than when it is at aphelion—farthest from the earth. The distance varies by 1.7 percent of the average. The earth is closest to the sun in January, with the result that we may then have the year's highest tides. It is farthest away in July, and that is when the sun does least to augment the tidal power of the moon.

Another factor is the declination of the moon and sun—their angular distance north or south of the equator. The earth's axis is tilted by 23½ degrees with respect to the plane of its orbit around the sun, and therefore as the earth makes its annual circuit the sun appears to move back and forth across the equator, bringing summer to the northern hemisphere when it is overhead and winter when it is below the equator. The moon also appears to move back and forth across the equator, for the plane of its orbit is tilted at an angle of five degrees to the earth's orbit. While the sun spends a year making its apparent north-and-south migration across the equator, the moon requires only 27⅓ days (the tropical month).

When the moon and sun happen to be in the same declination, there is an additional augmentation of their tidal forces. If this occurs at the time of the equinoxes, when the sun is crossing the equator on about March 21 and September 23, the sun can add 27 percent to the moon's tidal force, and we have *equinoctial tides*, higher than usual.

Regardless of the declination of the sun, when the moon is over the equator, the tides in the northern and southern hemispheres are of about equal heights because the moon is exerting equal force on both hemispheres. Such tides are known as *equatorial tides*. Two tides in a day are about the same.

34

The position of the moon above or below the equator determines the diurnal inequality of the tides. As the earth rotates northern and southern hemispheres are subject alternately to a greater or lesser tidal force, making a difference in height of morning and evening tides. *(Drawing by Richard L. Murphy, Jr.)*

But when the moon moves above or below the equator the tidal bulges it produces are askew. At the same time it is causing a maximum bulge for a tide in the northern hemisphere, it is causing a maximum bulge on the other side of the world in the southern hemisphere. As the earth rotates, the next high tides in those locations will be lesser ones. This effect is known as the *diurnal inequality* of the tides. It is most pronounced when the moon's greatest declination occurs at summer and winter solstices in June and December when the sun's declination is greatest. Such effects vary in different parts of the world.

Declination can also cause irregularity in the time between tides, changing it markedly from the average of twelve hours and twenty-five minutes. It can affect the time of tides with respect to the passage of the moon through its zenith. I have pointed out that except at, or near, syzygy (conjunction or opposition) the moon and sun are pulling at cross purposes; the height and time of a tide tend to be the products of a sort of compromise between them, with the moon usually exerting the greatest influence. The compromise may result in what is known as the *priming of the tides*, meaning that a high tide occurs before the moon reaches its zenith. Or there may be a *lagging of the tides*, when high tide occurs after the moon has reached its zenith. Another factor in these effects is that the waters do not respond with alacrity to gravitational pull. They are held back by inertia and friction.

The time between the passage of the moon through its zenith and the occurrence of the following high water or low water is fairly constant in any given location. It is known as the *lunitidal interval*, and given this figure for any port, one can come close to predicting the time of tides. The interval between full moon or new moon and the greatest spring tide that follows is known as the *age of the tide*. In the North Atlantic it is about one and a half days but in some places it may be as long as seven days.

For simplification (and perhaps also regional myopia) I have been discussing semidiurnal tides such as occur on the American and European shores of the Atlantic. But there are two other very important types of tides—diurnal, or once-daily, and mixed tides, which combine characteristics of diurnal and semidiurnal. An explanation of why they occur will come later.

Diurnal tides predominate in the Gulf of Mexico and are characteristic of some parts of the Pacific Ocean — the Philippines and some other islands and certain places in Alaska.

Tides in the Gulf of Mexico, unless driven by high winds, are very weak. The following table shows the heights of tides (above mean low water) for the first three days of September 1978 at Pensacola, Florida:

	HIGH	LOW
September 1	1.4 feet	0.6 feet
September 2	1.4	0.7
September 3	1.3	0.8

These were diurnal tides and, as can be seen, the range was not very great. But the next day the moon was over the equator, and for two days there were semidiurnal tides, also with a very small range:

September 4	0.9 feet	0.8 feet
	1.1	0.9
September 5	1.0	0.8
	1.0	0.9

The difference between low and high tide on these days was so slight as to be imperceptible to most Labor Day boaters. For the rest of the month, Pensacola returned to diurnal tides, with a mean range of only slightly more than a foot.

Mixed tides, combining characteristics of diurnal and semidiurnal, are found on the West Coast of the United States. There is a marked inequality in the heights of the succeeding tides — especially the low waters — and there is also an inequality in the time. But usually there are two high tides and two low tides each day. Typically there is a high tide, then a low tide, then a scanty high tide, then a moderate low tide. Here, for example, are the tides for June 13, 1979 at San Francisco's Golden Gate:

12:37 a.m.	6.1 feet
7:16 a.m.	−1.4
2:46 p.m.	5.0
7:27 p.m.	2.6

These pictures of a tide recording station at Anchorage, Alaska, show a tidal range of thirty-four feet. *(NOAA)*

feet. So there must be other factors beside the movements of the sun and moon to explain such variations (and many other variations as well). Indeed there are, and in the next chapter I will discuss some of them, first finding it necessary, however, to detour for a discussion of waves.

NOTES

The riddle by Aldhelm (c. 640–709), Bishop of Sherborne, was translated from the Latin by J. H. Pitman and appears in *Northumbria in the Days of Bede* by Peter Hunter Blair (St. Martin's Press, New York, 1976). It is reprinted by permission of the publisher.

In trying to explain the tides in simple language I have been aided by a large number of books, many of them not so simple, since a full explanation requires rather complex mathematics. I have already listed Darwin's *The Tides* and Marmer's *The Tide*. *The Tides* by Edward P. Clancy (Science Study Series, Doubleday, Garden City, 1968) and *Ebb and Flow* by Albert Defant (University of Michigan Press, Ann Arbor, 1958) are two of the best concise works for those with some technical sophistication. *Tides* by D. H. Macmillan (American Elsevier, New York, 1966) is comparable.

In periodicals, the best simple exposition I have found was in a two-part series in *Natural History* in 1959, Vol. LXVIII, "The Tides" by Thomas D. Nicholson (No. 6, June–July) and "The Margins of the Restless Ocean" by Hubert A. Bauer (No. 8, October). For the junior high school age, *The Rise and Fall of the Seas* by Ruth Brindze (Harcourt, Brace & World, New York, 1964) is acceptable. Bernard D. Zetler of the University of California, San Diego, sent me copies of his articles, "Tide Talk," which appeared in the magazine *Go* in Florida in 1968, succinct and helpful.

A large number of textbooks and other works in oceanography have appeared in recent years and many of them contain discussions of the tides. I found *The Seas in Motion* by F. G. Walton Smith (Thomas Y. Crowell, New York, 1973) particularly useful. Among others I have consulted are *Introduction to Oceanography* by David A. Ross (Appleton-Century-Crofts, New York, 1970); *Oceanography, A View of the Earth* by M. Grant Gross (Prentice-Hall, Englewood Cliffs, N.J., 1972); *The World Ocean* by William A. Anikouchine and Richard W. Sternberg (Prentice-Hall, 1973); *The Problem of the Tides* by E. C. Abendanon (C. Blommendaal, The Hague, 1960); *Man and the Sea*, edited by Bernard L. Gordon (Doubleday Natural History Press, Garden City, 1970).

Books that are particularly readable as well as informative include *Frontiers of the Sea* by Robert C. Cowen (Doubleday, Garden City, 1960) and *The Sea Around Us* by Rachel L. Carson (Oxford University Press, London, 1951, Mentor Books, New York, 1954).

Specific tide figures are from tide tables issued by the National Ocean Survey, National Oceanic and Atmospheric Administration, U.S. Department of Commerce. Other NOS publications of value include a booklet, *Our Restless Tides* (45 cents) and *Tide and Current Glossary* (75 cents) which may be obtained from the Superintendent of Documents, U.S. Government Printing Office. An excellent article by Steacy D. Hicks, chief of the Oceanography Division of the Coast and Geodetic Survey, "Ocean Tides," was published by *Pilot Chart*, U.S. Naval Oceanographic Office, in 1969.

The Bay of Fundy high record comes from an article, "The Towering Tides of Fundy" by Clyde M. Smith, which appeared in *National Wildlife* (Vol. 12, No. 2, February–March 1974). John James Audubon tells about measuring the Fundy tide in *Audubon and His Journals,* Vol. II (Dover, New York, 1960), in an account of his trip to the Maritime Provinces.

Waves, Seiches, and Amphidromes

"Let there be Light!"

"We have a bit of a problem," the Chief Engineer said. "The Chief Scientist believes we will need a way to transmit the light. He recommends oscillations. Waves. We'll have the same sort of a problem in the waters under the firmament."

"Very well. Let there be Waves!"

IF ON A CALM DAY when there are no power boats churning the water you look intently at the surface of a bay or lake, you will become aware of an intricate texture of waves. Cat's-paws of breezes, or perhaps fish jumping, send out ripples which criss-cross each other, are reflected by irregularities of the shore in different directions, merge, augment each other, or, if out of phase, combine to flatten the water. No eye can follow all the movement, but it is possible to see the changing patterns as one could never see them in a choppy or storm-tossed sea. The principles of wave motion are the same in both cases.

I have mentioned tide waves which, theoretically, would be 12,500 miles long, from crest to crest, as they were pulled around the earth at its greatest circumference—at the equator. These are what are known as *forced waves*, induced by the gravitational attraction of the sun and moon. Tidal movement also generates *free waves* which, like the ripples from a cat's paw, travel on and on after the force that produced them has disappeared. Both these kinds of waves are known as *progressive waves*.

Another kind of wave is the *standing, or stationary, wave*—a wave that is going nowhere. I have told how Galileo saw the water in Venice-bound barges slopping back and forth. What he saw was a standing wave. If you have tried to carry a large pan brimful of water, you have found that it inevitably spills because only a slight tilt makes the water oscillate—high at one end, then at the other. A full coffee cup is easier to carry because it is smaller and the oscillations are not as great.

Perhaps the easiest standing wave to understand is the *seiche* wave (pronounced as if spelled "saysh"). *Seiche* is a Swiss word first used at Lake Geneva, where the seiches are particularly noticeable and have long been studied. People were once mystified by the fact that, at times when there was no wind to cause waves, the water would rise and fall on the shore, as much as five feet. Yet tidal movement in the lake is negligible.

Careful observation showed that the lake sometimes oscillates in a standing wave, first high at one end and then at the other. The motion can be compared to that of a seesaw. The ends go up and down and there is no vertical movement at the fulcrum, which, in a wave, is called a *node*. If you can imagine a seesaw 100 yards long you will realize that the *period* of its movement (the time interval) would be much greater than the period of a 10-foot seesaw; and it would not change very much, no matter how hard the child at each end pushed with his feet when he touched the ground.

A dishpan and a coffee cup have different periods of oscillation. Every body of water has a characteristic period of oscillation, depending on its length, depth, and shape. Lake Geneva is about forty-four miles long; its average depth is 455 feet and, although it is curved, it can be regarded as an elongated basin. Calculations show that its period of oscillation is 74.4 minutes. One scientist observed a seiche through a whole week with a period of 74 minutes. In other words, it took about an hour and a quarter for the lake to seesaw. Lake Vättern in Sweden has also been studied carefully. It is seventy-seven miles long and the seiche period is 179 minutes.

The causes of seiches are varied and not thoroughly understood. A sudden squall or a change in barometric pressure at one end of a lake can start the rocking motion of the water, later to be observed at the other end of the lake where no disturbance was initially apparent. Earthquakes, even those too small to be noticed, can start vibrations of the water.

Harbors, open to the sea on one side, can have seiches caused by disturbances in the ocean. Willard Bascom, who made extensive studies of waves on the West Coast, reported that in Los Angeles Harbor, when the water appears perfectly calm, seiches can occur that lift ships as much as ten feet, breaking mooring lines and piles. Tidal movements regularly induce seiches there — short-period oscillations of as much as 1½ feet.

Understanding of seiches helps in the understanding of tides. Having used the seesaw for a demonstration, let us now think in terms of a child's rope swing. If the reader is lucky enough to have had a giant swing, suspended from a very high branch of a tree, he is also fortunate in being able to visualize the long period of such a swing. He knows that to push the swing, the push must be started with the forward sweep. Each push, given at the moment when the swing starts to move forward, will send the swing higher. A pair of children standing on the seat can pump alternately on the forward and back sweeps and send the swing very high indeed. If the ropes are shorter, the pumping or pushing must be done faster. Whatever the length, the pumpers and pushers must keep in time with the rhythm of the swing — its characteristic period. An attempt to push a swing in opposition to its direction will not only break the rhythm but knock the pusher to the ground.

The principle of a swing's behavior is that of *resonance* — a fundamental principle in physics which is involved in all kinds of things, such as the vibrations of musical instruments and the tuning of radio or television sets to receive electromagnetic waves of certain frequencies (certain periods of oscillation). It is also involved in seiches and tide waves. An organ pipe of a particular length produces sound waves of a particular frequency. A body of water is also tuned to resonate, with waves, in a particular way. If the period of the body is about the same as the period of the force, the wave motion will be maximized.

An interesting example may be found in the Great Lakes, in which there are seiches and also tides so small that most people do not even realize that they exist. Lake Michigan, which is 340 miles long, has a period of oscillation in close resonance with the moon. It responds to the moon's pull twice daily, as the oceans do, and it produces a spring tide as high as three inches at Chicago. Lake Superior, more than 400 miles long, has smaller tides — a spring range of about 2.3 inches. It must be remembered, of course, that depth and shape are factors, as well as size, in determining the period of a body of water.

The Baltic Sea, about 916 miles long, is shallow and its opening to the tidal influences of the North Sea is severely constricted at the Kattegat bottleneck. It has a period of 27.3 hours and is virtually tideless. So many rivers pour into it (except during the winter when they are frozen) that its most striking feature is the surface layer, chiefly of fresh water, which makes its level higher than that of the North Sea and causes an outflow into the Skagerrak.

How about a large body of water open to the sea at one end, such as the Bay of Fundy? It stretches in a northeasterly direction between Nova Scotia and New Brunswick, about 170 miles long, 50 miles wide at the entrance and 30 miles wide at the head, where it divides into two narrowing, horn-like bays. At Yarmouth, Nova Scotia, near the entrance to Fundy, the mean range is 11.5 feet; yet places near the head, as I have pointed out, have the highest tides in the world.

Some authorities give the Fundy tide as the prime example of a stationary wave, with the node at the mouth of the bay. They have estimated its resonance as near that of the lunar period. It takes high tide only forty-eight minutes to move from Grand Manan Island, in the mouth of the bay, to Amherst, at one of the tips, a speed which indicates a seesaw type of movement rather than a progressive wave. Docks stand high and dry at Truro and some other places at low tide. In Chignecto Bay, fishermen string gill nets on fifteen-foot poles and at low water pick the fish off as one picks apples off a tree. The story is told that pigs foraging on mud flats have been drowned in the rapidly moving flow of the tide. H. A. Marmer reported (though he did not vouch for the tale) that the pigs learned to station a sentry on a high bluff to warn them when the water was beginning to rise.

Spectacular though the Fundy tides are, authorities differ in explaining them. Donald R. F. Harleman, an M.I.T. hydro-dynamicist, takes issue with those who claim that the only cause of their height is resonance—that the tide occurs simply as a stationary wave. He feels that they fail to allow for the effect of friction. Resonance is an important factor but the main reason Fundy tides are so high, he believes, is the amplification by shoaling bottoms and narrowing shores. The tide wave coming in from the deep ocean strikes the broad continental shelf in the Gulf of Maine and is increased in height as it crosses the 100-fathom Georges Bank. It is forced into the narrowing and shallowing Bay of Fundy and there is further amplified; it is also increased in the narrow inlets—a case of funnels inside of funnel at the end of a very large funnel.

"If this occurred somewhere else and a one-foot tide was increased to two feet, it would not seem remarkable," Dr. Harleman says.

Alfred C. Redfield, at age eighty-eight the grand old man of the Woods Hole (Massachusetts) Oceanographic Institution, agrees on the importance of the continental shelf as a factor. He has just completed an analysis of tidal phenomena all along the northeast coast and observes that "In every case, the height of the tide is related to the distance to the edge of the continental shelf." He has written that "the tide in Long Island Sound, in which the maximum range is 7.4 feet, and that in the Bay of Fundy, in which the maximum range is about 34 feet, are augmented about equally in their passage up the embayments. The form of the basin has little effect on the augmentation."

Summer folk often wonder about the marked difference between tides on the north and south sides of Cape Cod. At Wellfleet, on Cape Cod Bay, the mean range is 10 feet, while at Chatham, on Nantucket Sound, it is only 3.2 feet. And at Siasconset, on the eastern shore of Nantucket Island, it is only 1.2 feet.

This kind of difference, Redfield says, "is explained by the consideration that in crossing the continental shelf the tide behaves as a co-oscillation reflected from the outer coast and consequently is greater the more distant the coast is from the deep water of the ocean. North of Cape Cod the ocean tide must traverse the Gulf of Maine and consequently is greater." The Gulf of Maine and the Bay of Fundy should be thought of as a single tidal system.

Co-oscillation? Simply put, this means that the tide wave coming from the deep ocean is in tune with, and augments, the tide wave generated over the continental shelf and, in this case, reflected from the shore. This effect of augmentation can be seen in dramatic form if you watch moderately heavy surf striking a sea wall. A wave is reflected by the wall and moves backward toward the sea. If the timing is just right, its peak coincides with the peak of another wave, which is then bigger when it hits the wall. If the peak of the reflected wave coincides with the trough of an oncoming wave, the height of both is lessened. The water tends to flatten out. Perfectly regular trains of waves are unusual, and the irregularity of most shores complicates wave behavior, but an observant shore-walker can detect the effect, which is known as *reflectance*.

These are wind-driven waves, of course. Tide waves comply with the same principle, though they are so long that one cannot see this behavior. Depending on their timing, they can be reflected by the continental shelf or the shore and can augment or lessen the height of the next tides, those following at intervals of about 12 hours and 50 minutes.

"Most authorities just mention reflectance and then drop the subject," said Redfield, "but it is extremely important in determining the character of tides."

Reflectance apparently is not a factor in the Chesapeake Bay, where the tides are in great contrast to those of the Bay of Fundy. For one thing, the Chesapeake is much closer to the edge of the continental shelf. The mean range at Cape Charles—at the entrance—is only 2.4 feet. The bay is shallow and the tide wave is slowed down by friction, so that it takes 15 hours to advance along the 180 miles from mouth to head. This means that two high tides can occur in different parts of the bay at the same time, with a stretch of low water for about fifty miles between them. The wave is purely progressive. By the time it reaches Baltimore, it has lost much of its energy, and the mean range there is only 1.1 feet. Delaware Bay also has a simple progressive wave but tides are higher, with a range that is almost the same at Cape May and Philadelphia—about 5.8 feet.

The open oceans must be thought of as being divided into basins if one is to account for the nature of their tides. On a small scale, the Mediterranean Sea behaves as if it were two seas, divided as it is by the boot of Italy and by Sicily. In neither of these basins is there a substantial tide. At Alexandria the range is less than 4½ inches and at Naples, in the western Mediterranean, it is less than 6 inches. No wonder the Greeks and other classical seafarers paid little attention to the tides. In the Adriatic Sea, however, as I shall explain later, there is an important resonance effect.

Not enough records have been obtained to be certain about tides in the open oceans, but theories have been developed to account for their characteristics. The Pacific is thought to have five principal basins, in each of which enormous standing waves rock back and forth in response to tidal forces. Each of these stationary waves has a node at which there is very little vertical movement of the sea.

Tahiti, far out in the middle of the South Pacific, is believed to be at the node of such a wave. Although the moon above the palm trees provides romantic nights, it has little gravitational influence on the tropical waters that sparkle with moonbeams.

Tahiti has two high tides (and not very high ones) each day—one usually at noon when the sun is overhead and one at midnight when the sun is on the opposite side of the world. The native word for midnight is the same as the word for high tide. Obviously the sun, not the moon, governs the rhythm of the sea in the Society Islands. The moon is out of step with the stationary wave that it has created elsewhere.

The head of another stationary wave of the Pacific (a relatively small one) is in the Gulf of Panama. Because of this, the mean range at Balboa, at the Pacific end of the Panama Canal, is 12.6 feet. At Cristobal, at the Atlantic end of the canal, the mean range is only 0.7 feet. Mean sea level at Balboa is nine inches higher than at Cristobal and this difference, plus the tidal difference, creates a difficult problem to solve in the scheme to build a sea-level canal between the Atlantic and the Pacific.

But why should Cristobal have such a scanty tide? The answer, put simply, is that there is a nodal point in the Caribbean for a stationary wave in the Atlantic. Puerto Rico has a mean tidal range of only 1.5 inches—barely enough to wet the toes of a sunbather.

The modesty of tides in the Caribbean, plus such factors as island barriers and sea bottom irregularities, is related to peculiarities of tides in the Gulf of Mexico. Naples, on the west coast of Florida, is only 125 land miles from Miami on the east coast, yet its tides come eight hours later, apparently because it takes the tide wave so long to move through the narrow Straits of Florida. The range is about the same, 2.5 feet at Miami and 2.1 at Naples. Although Naples and the shore below it have two tides a day, the tendency toward diurnal tides begins to be felt there; and when the moon is in extreme declination, Naples has diurnal tides similar to the usual ones at Pensacola and points west. At New Orleans the flow of the Mississippi River is so great that there is no perceptible tide when the river is high and a range of not more than 0.8 of a foot when the river is in its low stages.

To understand why tides in the Gulf of Mexico tend to be diurnal and those on the Pacific Coast are a mixture of diurnal and semidiurnal, one needs to recall the remark in the first paragraph of this chapter that merging waves can augment each other or, if

50

out of phase, combine to flatten the water. I have been writing of the lunar and solar tidal forces as if there were only two at work, but actually there are a number of constituents (as analyzed by scientists) and each constituent can be thought of as producing a wave. The way in which all these waves react with each other—along with such factors as resonance, declination and the shape of shores—determines the kind of tides that will occur in a given location. The interplay of the constituents can be extremely complicated.

Diurnal inequality is particularly influential on mixed tides. When the moon is over the equator, the difference between the twice-daily high tides or twice-daily low tides may be relatively small but the differences become great when the moon's declination increases. In Los Angeles Harbor, for instance, a lower high water and a succeeding higher low water can be about the same height, producing what is known as a *vanishing tide*, when the water stands at the same level for some three hours.

Another complication in the character of tides is *amphidromic movement*. The Greeks have an ancient custom—the Amphidromia—which is performed at the christening of a baby: the child is carried around a circle of relatives and friends and introduced to them by name. In many places the tides circle in an amphidromic way. To make this movement graphic, oceanographers draw cotidal lines, one set connecting the points at which tides occur at the same time and another set connecting the points at which the tides reach the same height.

The cause of amphidromes is the *gyroscopic*, or *Coriolis, force*, named for Gaspard G. de Coriolis, a French mathematician of the early nineteenth century. You have experienced the Coriolis effect if, on a merry-go-round, you have tried to walk in a straight line inward or outward—at a right angle to the direction of rotation. You have felt yourself being deflected. So it is on the rotating earth. Tides in the northern hemisphere are deflected toward the right (counterclockwise) and those in the southern hemisphere, toward the left (clockwise). At the equator the force is zero. (The rotation of hurricanes and cyclones is affected in the same way.) There are exceptions to the rule.

Tides in the seas around the British Isles provide an interesting example of how the Coriolis force works. As the tide wave moves northward in the Atlantic through the English Channel, the water is deflected toward the French coast, where high tides are higher than on the southern shore of England. Mont-Saint-Michel gets tides of more than forty feet. But the tide wave does not move northward all the way through the Channel, and as the ebb begins, the sea — still sloshing toward the right — causes low tide to be higher on the English shore than low tide on the French coast.

To reach the North Sea, the tide moves eastward above the tip of Scotland and then southward along the Scottish and English shore. It then circles past the coast of Holland, arriving at Denmark some fifteen hours after it entered the North Sea. The rotary movement is around a node east of Denmark where there is practically no tide. Co-range lines are also circular, but while the mean range on the English coast is as high as sixteen feet, it is only three feet or so on the Danish coast. To complicate matters, there is another node at the tip of Norway, with an amphidromic movement around it. Here the tide wave moves more rapidly and the tidal range near Stavanger is ten inches or less, due, no doubt, to the shallowness of the water.

Mariners bound north through the English Channel have long known that if they ride a fair tide toward Dover they can then move into the amphidromic system of the North Sea and continue to be helped by a northward current.

Curious tides occur in the Solent, the arm of the English Channel that runs behind the Isle of Wight. In the eighth century the Venerable Bede wrote: "In this sea the two ocean tides which break upon Britain from the boundless northern ocean meet daily in conflict beyond the mouth of the river Hamble. . . . When their conflict is over they flow back into the ocean whence they came." The result of this conflict is that at Southhampton high tide remains for an unusually long time; under some conditions, the tide may fall slightly and then rise again before it ebbs for low water, creating a double high tide. Presumably the cause is a difference in the movement of tide waves around the Isle of Wight.

Returning to the subject of amphidromic systems, I should

point out that a very important one exists in the North Atlantic west of the British Isles. A cotidal chart of that area has the look of a web spun by a drunken spider. The tide circles, counterclockwise, around a node east of Newfoundland, touching first Iceland, then Greenland, then Labrador and Nova Scotia. Albert Defant, one of the authorities on tides, writes in *Ebb and Flow* that the tide wave "loses a great deal of energy in crossing the barriers separating it from the Arctic Ocean, and later in crossing the ice of the polar region." The reflected wave, he believes, is weak and "The resonance tide, instead of having the form of a simple stationary wave, must be thought of as a set of superposed waves that are out of phase."

In discussing stationary waves in the Pacific, I deliberately deferred mentioning amphidromic effects. Amphidromes exist there as well as in the Atlantic. Stationary and progressive waves revolve in great circles but the Pacific is so vast that the tides are not fully understood. For that matter, knowledge of tides in the Atlantic is not complete.

I referred to the node at Tahiti, with tides at noon and midnight. This node apparently is the center of an amphidrome which rotates in a counterclockwise direction around the Society Islands. Another one is centered at a point about half-way between North America and the Hawaiian Islands. There seems to be a complex of three nodes in the Western Pacific, centered in the Solomon Islands, the Caroline Islands, and Bonin Island (which is farther north, toward Japan).

Tide measurements are made at many islands in the Pacific, of course, but these represent only a small part of the ocean. In recent years new techniques have been developed for measuring tides in the open ocean, using radar and laser ranging from satellites and buoys; these instruments determine the pressure below the surface of the water (and thus measure the height of tide waves above) and telemeter the data to oceanographers. Generally speaking, the tidal range in the open oceans is only a foot or two.

Internal tide waves—that is, tides beneath the surface of the ocean—have also been detected. This detection is possible

V

High Winds, High Tides

Nature bats last. PAUL GODFREY

"WORST STORM of century," proclaimed a newspaper's banner headline. No one could argue very hard to the contrary about the "Blizzard of '78," which destroyed or damaged more than 7000 homes in New England and took ninety-nine lives.

Inevitably comparisons were made with other great storms, in which more lives were lost or more ships wrecked. Three indisputable records at Boston were established—the most snow in twenty-four hours (23.7 inches); the most snow in a single storm (27.1 inches) and the most snow on the ground (29 inches).

It is not with snow that we are concerned here, however, although snow was important to the passengers of thousands of cars stranded in deep drifts (some died there) and to millions of people who were snowbound for a week (many were without heat or suffered other hardships). What is of more interest to this book is the disastrous effect of storm-driven tides. Portland, Maine, had the highest tide on record and, although Boston lacks full documentation, the tides that flooded Massachusetts shores were probably the highest in history. From New Jersey to Maine, 7858 families reported losses, according to the American Red Cross, and nearly all of those were from tide-flooded homes. In Massachusetts, where the devastation was greatest, an estimated 10,000 people were evacuated from the shore.

The full significance of tidal flooding has received little attention from the public or from weather experts.

The Blizzard of '78 occurred on February 6–7 at a time when the moon was especially close to the earth and it was in syzygy with the sun. Nearly forgotten at that time was that another violent storm, though less disastrous and spectacular, had occurred under similar circumstances on January 8–9. And though East Coast people, preoccupied with their own problems, were hardly aware of it, the West Coast also had tempests with severe coastal flooding at both times.

Two other dates should be noted here, with an explanation to come later in this chapter: they are January 8 and February 7, 1974.

The first comprehensive study of catastrophes that result from a combination of high winds and extraordinarily high tides has recently been completed for the National Ocean Survey by Fergus J. Wood, an authority on tides and weather; this study has important implications for everyone who lives on or near the edge of the ocean. Whether there is a cause-and-effect relationship between coastal storms and tide-producing forces no one can yet say, but they make a malevolent alliance.

As a research associate for NOS (the Federal agency which watches over the oceans), Wood became intrigued with the problem while reviewing the vicious storm of March 6–7, 1962, which lashed the Atlantic shore from Maine to South Carolina, taking forty lives and causing a half-billion dollars' damage. Snow and northeasterly winds of thirty-five to forty-five miles per hour were forecast for the mid-Atlantic shore, but as it turned out there were howling onshore winds with gusts up to eighty-four miles per hour, and they persisted for about sixty-five hours. Giant waves

High tide at Hingham, Massachusetts, February 7, 1978. High water, predicted at 11.5 feet, was accompanied by high winds and low barometric pressure, which caused the sea to wash well over the pier at the Hingham Yacht Club, pictured, doing severe damage. *(Photo by author)*

and swell washed away houses and drowned people unfortunate enough to be caught on low land.

The moon was new and a spring tide of 5.2 feet was predicted for Atlantic City. Actually it reached 9 feet and there were five successive abnormally high tides, remembered by shore dwellers as the "High Five."

Analysis of the weather showed what caused the storm. Two low-pressure centers, one from the west and one from the southeast, converged at Cape Hatteras. High pressure areas, one from Labrador and one from the south, squeezed and blocked the augmented low and kept it from moving out to sea. A drop of one inch in barometric pressure allows the sea (acting like the pool of mercury in a barometer) to rise 13.2 inches; thus the low pressure during the storm was one reason for the high tides. Onshore winds, piling up the water as they blew toward the shore, drove the sea even higher.

But Wood, looking at the data a dozen years later, was not content with the common explanation of the extreme coastal flooding as due to "a coincidence of spring tides and high winds." He had a hunch that additional and more complex tide-raising factors were involved in the disaster. Examining the astronomical record, he found that indeed certain tide-amplifying conditions had been overlooked. Not only were the moon and sun in syzygy, which produces spring tides twice a month, but the moon was in perigee just thirty-one minutes before syzygy. Furthermore, the moon was considerably closer to the earth than it usually is at perigee. These circumstances produced an extremely high tide.

Fergus Wood has combined long study and experience in astronomy, physics, and meteorology. Born in Ontario, he was a student at Berkeley in 1937 when an unusually close and fast-moving asteroid, Hermes, 1½ miles in diameter, was discovered photographically in Europe. Initially it was found to be rushing toward the earth with a velocity which, if continued, would have caused it to crash within 6½ hours. Overnight, and during the next day, Wood helped compute an orbit which eventually showed that the body had swerved and was heading away from us. ("The hardest work I ever did in my life," he says.)

Wood was an Air Force general-staff meteorologist during World War II, and later, a scientist at The Johns Hopkins University Applied Physics Laboratory, at the University of Maryland, and for NASA before he joined the National Ocean Survey in Washington, D.C. For the U.S. Navy, he prepared a detailed technical monograph on wind influences over capital ships at sea. He edited the U.S. Coast and Geodetic Survey's three-volume report on the Alaskan earthquake of 1964 — one of the most comprehensive studies ever made of an earthquake. Now 61, he has an almost boyish enthusiasm for science. Of the vicious winter storm of 1962, Wood has said, "It struck me as unforgivable that we would have such a disaster without a better warning. Reliable weather forecasts can, with the advent of the weather satellite, now be made several days in advance, but in the past, at least, no adequate attention has been given to the tides. And there was something special about the tides during the 1962 storm."

As I have pointed out, syzygy and lunar perigee were close together at the time of the 1962 storm. This meant that the moon and sun were lined up, exerting a combined gravitational attraction, and the moon, being at the position in its orbit closest to the earth, was attracting the sea even more than usual. (Remember, tidal force varies inversely with the cube of the distance between two bodies.) A tide produced by such a situation is known as a *perigean spring tide* and its range may be forty percent greater than average. Such *ordinary* perigean spring tides occur twice each year.

But there can be a situation — occurring, on the average, not oftener than once every one and one-half years — when even higher tides can be raised. Each time the moon comes into perigee–syzygy alignment, the sun distorts the lunar orbit and makes the moon move even nearer to the earth. The closer the alignment, the closer the moon comes to the earth. The smaller the lunar distance from the earth, the greater the tide. Wood coined the term *proxigean spring tide* for this case of especially large perigean spring tides, those that occur when the moon at perigee is unusually near the earth. If there happens to be a strong onshore wind at the same time, as in the 1962 storm, coastal flooding is inevitable in lowland regions subject to considerable lunitidal ranges.

Another factor that has often been ignored, Wood found, is that as the moon swings around the perigean portion of its orbit it travels faster. A point on the rotating earth must then take longer to catch up with the moon and the tidal day is lengthened. The increase in lunar speed, through its resulting prolongation of the amplified tide-raising forces, makes the tides rise higher. They also rise more rapidly, remain longer at or above the critical level for flooding, and flood farther inland. And the farther inland the tide intrudes — filling all sorts of inlets, marshes, and basins that it usually does not reach — the longer it takes to ebb.

A strong offshore wind and high barometric pressure at the time of a perigean or proxigean spring tide can, as contrasted with the above situation, cause an extremely low tide in the low-water phase.

Suspecting that many cases of storm disaster could be explained in part by perigean or proxigean spring tides, Wood worked with the U.S. Naval Observatory on a computer printout showing all the dates from 1600 to 1999 on which syzygy and the moon's perigee occurred or are scheduled to occur close together. Then he laboriously checked through every available source to determine what the weather and tides were like on the past dates. Official U.S. records go back only about a century but he was able to dig much of the information he wanted from old newspaper accounts of coastal flooding and from historical documents. As a representative group, he researched one hundred cases of major coastal flooding when syzygy and perigee were in proximity.

The earliest example in North American waters that Wood found was the famous tempest of 1635 which Governor William Bradford described in his *History of Plimoth Plantation* as occurring on August 14-15 (Old Style Calendar):

> . . . such a mighty storme of wind and raine
> as none living in these parts, either English or
> Indiens, ever saw. Being like (for the time it

continued) to those Hauricanes and Tuffons
that writers make mention of in the Indeas.

It began in the morning, a litle before day,
and grue not by degrees, but came with a
violence in the beginning, to the great
amasmente of many. It blew downe sundry
[211] houses & uncovered others; diverce vessels
were lost at sea, and many more in extreme
danger. It caused the sea to swell (to the
southward of this place) above 20 foote, right
up & downe, and made many of the Indeans to
clime into trees for their saftie. . . . The signes
and marks of it will remaine this 100 years in
these parts wher it was sorest. The moone suffered a great eclips the 2. night after it.

Modern authorities believe the storm was indeed a "tuffon"
(typhoon) and Wood lists it not among his one hundred cases but
among thirty representative hurricanes "occurring nearly concurrently with perigean spring tides."

Syzygy came two nights after the storm when the moon, earth,
and sun were so perfectly lined up that there was a lunar eclipse.
Perigee was less than forty-two hours after syzygy. This combination of circumstances was enough to have substantially augmented the tides. Wood observes, however, that "a fast-moving
hurricane does not usually provide as much time for a buildup of
water level by friction at the air-sea interface as does a stagnant,
offshore extratropical storm possessing a long overwater wind
path."

Information about the 1635 hurricane is sparse but Wood
thinks it may have taken on some of the extratropical characteristics that were observed in the deadly 1938 hurricane that hit
southern New England.

More nearly typical of the proxigean spring tides was the storm
of March 7, 1723 (New Style Calendar), about which *The Boston
News–Letter* reported:

Yesterday, being the Lord's Day, the Water
flowed over the Wharff's and into our Streets to
a very surprizing height. They say the Tide rose

Fergus J. Wood, now retired from the National Ocean Survey, still keeps an eye on perigean spring tides in Southern California. *(Photo by Lt (jg.) Leland H. Ward, USN)*

20 Inches higher than ever known before. The
Storm was very strong at North-east. . . .

The loss and damage sustained is very great,
and the little Image of an Inundation which we
had, look'd very dreadful . . . the tide rising to
a height of 16 ft. . . . At Hampton, New
Hampshire, the storm caused the great waves of
the full sea to break over its natural banks for
miles altogether, and the ocean continued to
pour its water over them for several hours.

No less an authority than the Rev. Cotton Mather wrote to the
Royal Society about this storm:

. . . our American philosophers observed an un-
common concurrence of all those causes which a
high tide was to be expected from. The moon
was then at the change, and both sun and moon
together on the meridian. The moon was in her
perigee, and the sun was near to his having past
it, but a little before. Both the sun and moon
were near the Equinoctial, and so fell within the
annual and diurnal motion of the terraqueous
globe. There was a great fall of snow and rain,
the temper of the air was cool and moist, and
such as contributed unto a mighty descent of
vapours. A cloudy atmosphere might also help
something to swell and raise the waters. Finally
the wind was high, and blew hard and long,
first from the southward, and it threw the
southern sea in a vast quantity to the northern
shores: Then veering eastwardly, it brought the
eastern seas almost upon them. And then still
veering to the northward, it brought them all
with even more accumulations upon us. They
raised the tide unto a height which had never
been seen in the memory of man among us. . . .
It filled all the cellars, and filled the floors of the
lower rooms in the houses and warehouses of the
town. The damage was inexpressible in the
country. On the inside of Cape Cod the tide rose
four feet, and without, it rose ten or a dozen feet
higher than was ever known.

Cotton Mather was shrewd in observing the significance of lunar and solar positions. Syzygy and perigee came just six hours apart, on the day before the storm. Fergus Wood notes:

> On the east coast of the United States, the normal lag between the occurrence of such a combined astronomical event and the resulting perigean spring tides produced is approximately 1 to 1½ days.

The 1723 storm was almost exactly on the button.

Among the other cases that Wood analyzed there were two of special interest in 1846. (It is possible for three to five perigean spring tides in which the separation between perigee and syzygy is less than twenty-four hours to occur in one year.) In the first case, a storm on March 1, three days after perigee–syzygy at new moon, started to breach the Hatteras Outer Banks which protect Pamlico Sound in North Carolina. In the second, on September 7-8 a storm just one day after the maximum effect of perigee–syzygy at full moon completed the job, creating a new, navigable inlet through the barrier. A third storm at a closer perigee–syzygy alignment in October scoured it enough to permit passage of ships.

Fifteen years later, when the Civil War began, the Confederacy used the new inlet for blockade runners and for privateers to ambush passing Union ships. Captured Union sea captains escaped and carried back information about Confederate operations through the channel. In August 1861 the Union Navy sent an expedition through the inlet, subdued Fort Hatteras and Fort Clark, and established a strong base in Pamlico Sound.

Headlines of newspaper clippings that Wood has collected tell of the often devastating consequences of the perigean spring tides, although the full explanation of the cause of the associated damage was not known when they appeared. For example, from *The Boston Herald* of November 25, 1885:

A MIGHTY TIDE

— — —

Old Neptune Baptizes
the Shore

— — —

An Unprecedented
Rise of Water

— — —

Picturesque Commingling
of Wind and Wave

— — —

Great Damage to Property
in New York

— — —

The Jersey Coast Strewn
With Wreckage

In the Boston area, the *Herald* reported, the "briny elements" washed over sea walls and did widespread damage.

From the *Los Angeles Times,* December 18, 1914:

SEAS LASHED BY GALE

BATTER COAST TOWNS

Houses Destroyed, Bulkheads Shattered, Sewer and Gas Mains Severed by Pounding Breakers on Crest of High Tide — More Trouble Feared Today — Loss of Property Many Thousands — No Casualties

The New York Times, March 6, 1931, ran this story on Boston:

THIRD GREAT TIDE

LASHES BAY STATE

Towering seas continued to lash the coast of New England early today despite the fact that the wind and snow storm which accompanied yesterday's record-breaking tides had moved off-shore. . . . The waves of the third consecutive abnormal tide, though somewhat abated, swept in at noon today and toppled several beach houses which had been weakened by the previous more savage onslaughts.

The loss is expected to run into the millions . . . great swells broke over seawalls an hour before high tide.

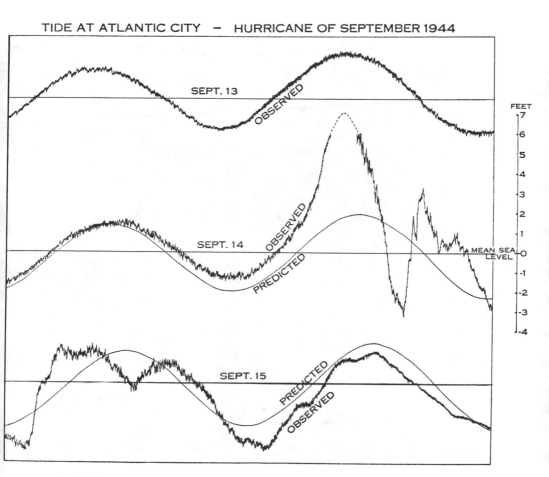

TIDE AT ATLANTIC CITY — HURRICANE OF SEPTEMBER 1944

Graphs show the effect of hurricane winds on tides at Atlantic City, New Jersey in September 1944. Two-foot tides were predicted for the fourteenth; but when the hurricane hit, the sea rose to nearly seven feet. *(NOAA)*

and Dutch coastal areas, disrupting shipping
and sending waves pouring over dikes at Ostend
and other points. At high tide the Thames rose
to just 18 inches below the level which would
have sent floodwaters into the city.

A month later, perigee preceded by forty-two hours the syzygy at new moon, which occurred at 9:54 P.M. Eastern Standard Time on February 7. This separation qualified it for Wood's designation of *pseudo-perigean spring tide*. (In the same class was one of New England's most famous storms, the Minot's Light Storm of April 14–16, 1851, so called because it dramatically toppled a lighthouse off Scituate, Massachusetts. The tide was so high that it made Boston an island, and caused heavy damage by flooding all along the coast.)

Emphasis in the broadcast forecasts for Monday, February 6, 1978, was on the prediction of heavy snow. However, the marine forecast for Boston Harbor and adjacent waters predicted that a nor'easter would bring "tides two to three feet above normal tonight and tomorrow. Rough surf and some beach erosion on east-facing shorelines. Some flooding of low-lying areas at time of high tide tonight and tomorrow." That proved to be an understatement.

Snow began falling about noon and by dusk was not merely falling—it was being driven horizontally by gale winds, to come to earth in the lee of buildings and pile up in drifts that made streets and highways impassable. A blinding, bone-chilling blizzard had set in and would continue for thirty-six hours. (Technically, it was not a blizzard because the temperature was above twenty degrees, but everybody called it one.)

A high tide of 10.2 feet had been predicted for 10:20 P.M., but three hours in advance tremendous seas, driven by northeast winds that peaked at eighty miles per hour (at 9 P.M.), were washing across low shores north and south of Boston where there were thousands of houses, most of them only a few yards apart, built by people who wanted to be close to the beaches. Residents had experienced tidal flooding of streets before and supposed that if they just sat tight the water would go away. Instead, it got deeper, rushing into houses and carrying eight-inch thick cakes of ice which had previously been formed on beaches. Residents retreated to second floors and attics.

Massive concrete sea walls, which in the past had protected homes from surf, were pounded to pieces by the roaring waves. Houses were showered with tons of gravel and even by boulders thrown up from the ocean bottom. The breaking waves splintered homes, washed them off their foundations, knocked them against each other, tossed them into streets upside down.

Electrical connections short-circuited and started fires which, fanned by fierce winds, burned without restraint. Fire trucks from inland stations were blocked by deep snowdrifts or, if they managed to reach shore communities, by deep rockdrifts and high water. Firemen, policemen, and others waded neck-deep in the seas trying to rescue whole families marooned in their homes.

At Lighthouse Point in Scituate, on the South Shore, Herbert Fulton, a fire captain from nearby Norwell, and Brian McGowan, a volunteer, were trying to take four people to safety in an aluminum boat at about 1:00 A.M. through a street across which water ten feet deep was coursing. The tide was going out and suddenly a sea wall broke from the pressure of the water inside. The water gushed out through the gap, pulling the boat toward the sea. Fulton stuck his pike into a telephone pole in an attempt to stop the boat and McGowan caught an oar on a submerged automobile, breaking the oarlock.

The boat upset, dumping the passengers; a five-year-old girl and a sixty-three-year-old man were swept away and drowned. Fulton managed to pull the man's wife to the intact part of the sea wall and the child's mother and McGowan also struggled to safety, ice forming on their bodies as they emerged from the water.

Ebb tide offered some respite in the early hours of February 7, but with an even higher tide of 11.5 feet predicted for 10:39 A.M. and little abatement of the wind, flooding became worse. The Boston tide gauge showed 14.94 feet, but that was in sheltered water. On the shore, where giant waves were breaking over houses, an official figure was irrelevant.

At Nahant, when seventy-one-year-old John J. Donahue discovered water pouring through the cellar windows of his house, he tried to awaken the tenant who had an apartment in his basement. Suddenly there was an eight-foot surge of water, and the basement was filled all the way to the ceiling. Donahue and his wife were rescued by a front-end loader which was taking people from second-floor windows in its bucket. Three hours later the house caught fire and burned to the waterline. The tenant's body was found under the wreckage in the still-flooded basement three days later.

Such harrowing events were taking place all along the coast. A list of twenty-nine dead was compiled by *The Boston Globe,* but it included elderly men who died of heart attacks while shoveling snow and people asphyxiated by carbon monoxide in snow-bound cars or in kitchens where they were using gas ranges to keep warm. Five men from Gloucester were lost during the night in the *Can Do,* a boat in which they were trying to go to the aid of a Greek oil tanker that had gone aground in Salem Sound and was taking water in its engine room. Three Coast Guard boats had been unable to reach the tanker and one of them had grounded as well.

At Rockport, "Motif No. 1," a red lobster shack famous because it was a favorite subject for artists, was washed from its ancient granite pier into the harbor. At Nauset Beach on Cape Cod, "The Outermost House," a shrine because Henry Beston lived there while creating a book by that title, was blown into the waves from the high dune on which it stood.

On the West Coast, death and devastation were not as great; but gale winds built up higher-than-usual tides which swept over sandbag barriers in California at Surfside, Sunset, and Seal Beaches. The tides caused flooding on Balboa Peninsula, Balboa Island, and Pacific Beach and produced severe erosion at South Mission Beach.

After the New England catastrophe, in which eight Massachusetts counties had been declared a disaster area by the White House and federal troops had been sent in for rescue work, people began to pick up the pieces, file insurance claims and decide what to do next. The 1968 National Flood Insurance Program offered some relief. The Program makes low-cost insurance available where local governments restrict development on high-hazard lands. It can pay for the removal of damaged structures and turn the land over to municipalities for recreational purposes. Because of stricter wetlands regulations passed in recent years, some storm-damaged houses could not be rebuilt. Many shore dwellers wanted bigger and better sea walls. The state was urged to acquire strips of shore; there has been a chronic complaint that private owners should not monopolize beaches anyhow — that all citizens should have a right to swim there. Generally speaking, however, houses were restored and remain vulnerable to future catastrophes.

Professor Paul Godfrey, an ecologist from the University of Massachusetts, was conducting a study of coastal barrier islands. He visited Hull, a town that occupies a six-mile-long sandspit that connects drumlins and rocky promontories, at the entrance of Boston Harbor. Nantasket Beach, the largest beach in the area, once made Hull a resort town but in recent decades many people began living there year-round. Only a few hundred feet wide in places, the town was awash during the storm, with 3,858 houses damaged and 110 destroyed. The recently cleared thirty-three-acre urban renewal site on which promoters had wanted to build high-rise, Miami-style apartment buildings, was under five feet of water. A new sewage treatment plant (Hull's first major attempt to solve serious pollution problems) was damaged.

Briefed on plans for better sea walls at Hull, Godfrey observed that they often create as many problems as they solve and may encourage unwise real estate development. He pointed out that the sea level is rising and that the future will bring more great storms.

"Nature bats last," Godfrey said.

In preparing his table of perigee–syzygy situations in four centuries, Fergus Wood listed eighty-four in the years 1979 through 1999. Of these, he identified fourteen as *proxigean spring tides*, the first scheduled for January 28, 1979. If you live on the shore you will do well to paint the high hazard dates of the next decade on the lintel of your door and watch the weather — for

75

proxigean tides can produce severe coastal flooding if accompanied by high onshore winds. The dates are:

October 23, 1980
December 10, 1981
February 17, 1984
November 12, 1985
December 30, 1986
October 14, 1989

And then comes a series of three *extreme proxigean spring tides:*

December 2, 1990
January 19, 1992
March 8, 1993

As to whether strong onshore winds will occur within several days of the unusually high tides of those dates, Wood of course cannot predict. "Well, what are the odds?" I asked. "You have found a hundred cases in which high tides and high winds came together and there must be others that you haven't identified. According to folklore and in the opinion of some scientists, the moon has an influence on the weather. Is it possible that tides in the atmosphere, for instance, as well as in the sea may increase the likelihood of storms?"

"It's a tantalizing thought," Wood replied slowly. "I can only emphasize that from a scientific point of view we should learn more about atmospheric tides."

No one knows better than Wood that onshore storms can occur without simultaneous perigean spring tides, and vice versa. He has studied twenty-two cases in which ordinary spring tides coupled with onshore winds were associated with coastal flooding. Since such tides can occur twenty-five times in a calendar year, the possibility that they would coincide with storms is very high. But the coincidence of ordinary spring tides and coastal flooding is far less than between perigean spring tides and severe coastline inundation. Wood has written:

> The seemingly above-average frequency of such concurrent events raises the question whether some possible interrelationship be-

tween respective astronomical (gravitational) and meteorological phenomena might exist which has not been established. From the available, documented occurrences, a certain statistical relationship also seems to hold between the most severe cases of tidal flooding and the second or third alignment of a given perigee-syzygy series. Under these latter circumstances also, repeated flooding events often occur within consecutive anomalistic months.

When he wrote that, Wood was acutely aware of the storms of January 8-9 and February 6-7, 1978, and also of the West Coast storm of March 5 (at the time of perigee, 81.5 hours before syzygy) in which raging, high-tide seas caused heavy damage at Malibu Beach in California; the East Coast missed this one.

At the end of 1977, Wood had retired from NOS to live at San Diego; there he wrote a supplement, after the winter storms, to his magnum opus, a massive volume published by the U.S. Government Printing Office under the massive title *The Strategic Role of Perigean Spring Tides in Nautical History and North American Coastal Flooding, 1635-1976.* It was the first definitive study of the subject and probably the most important scientific work on the tides for the U.S. government since Rollin Arthur Harris confirmed the existence of amphidromes in 1904 and published a *Manual of Tides* (1894-1907) — a book that did not evaluate the combined effects of perigee and syzygy.

Fergus Wood gave less attention to hurricanes than to winter storms because the former have been more extensively studied and they can be sufficiently evil without the help of perigean spring tides. And, as has been pointed out, hurricanes move too fast to equal winter storms in duration of flooding. Case histories show that when a hurricane arrives at a time of low tide the principal damage is caused by wind. The higher the astronomical tide, the greater is the damage from coastal flooding.

Of the thirty American hurricanes which have arrived "nearly concurrently" with perigean spring tides and produced coastal flooding, I have mentioned two — in 1635 and 1938 — as being of similar character in that they were not purely tropical. The 1938 hurricane struck southern New England on September 21 at about the time of high tide, the day after the moon was in perigee and two days before syzygy at new moon.

Technically, the storm was not a hurricane when it first hit New York in the late afternoon. The wind reached only 65 miles an hour, and for a hurricane the wind must blow at least 74. Manhattan was practically paralyzed, however, and when the storm moved on to Long Island, the wind registered 110 miles an hour. Later the Blue Hill Observatory near Boston recorded gusts up to 186.

Narragansett Bay and Providence, at the head of the bay, were a central target for the hurricane. The water rose thirteen feet above the predicted level at Providence and coursed through the streets of the business section, eight feet deep. New Bedford, Massachusetts, and other shore areas were also hard hit.

Altogether, the 1938 hurricane took 680 lives and caused $400 million in property damage. It destroyed 4,500 houses and farm buildings and 2608 boats, with damage to many more; it uprooted or splintered 275 million trees and killed a half-million chickens. It was the first severe hurricane to strike New England in modern times and the peril had not been fully appreciated in advance. Possibly most of the deaths could have been avoided if there had been ample warning and people had realized the danger. As a consequence of the disaster, hurricane forecasting and warning were improved and barriers were built to give Providence and New Bedford greater protection.

The greatest loss of lives in American hurricane history — 6000 — was at Galveston, Texas, on September 8, 1900, by coincidence the day of a perigee-syzygy alignment. Wood assumes that this was merely a coincidence, for the range of tides at Galveston is too small for the sun and moon to have made much difference. The basic problem was that the boom city had been built on a sandbar only nine feet above sea level and a mile wide. The mean tide range there is only 1.4 feet, but high winds and low pressure pushed the waters to twenty feet or more. Houses were washed from their foundations and pounded each other to pieces. After that disaster, Galveston was rebuilt seventeen feet higher and an eight-mile sea wall was constructed. But in August 1915 another hurricane struck and drowned 275 people, with a property loss of $50,000.

A close second in the number of fatalities was the hurricane of September 16–17, 1928, which killed perhaps 2,200 in the West Indies and then close to 2500 in south central Florida. There have been many other such hurricanes, and it is believed that nine out of ten people killed have lost their lives in the surges of the sea. More people in the United States are endangered today than ever before because increasing numbers have chosen to live on low shores.

By far the most lethal of storms have been the tropical cyclones (the same as hurricanes) that form in the Indian Ocean and whirl up the Bay of Bengal to inundate the low-lying delta of the Ganges and Brahmaputra rivers, an indispensable granary teeming with millions of people. The mean range of tides in the Hooghly River, one of the many mouths of the Ganges, is 10.6 feet at Calcutta. In 1737, cyclone-driven waves said to have been forty feet high killed an estimated 300,000 people at Calcutta.

There have been a number of such disasters, but the worst was on November 13, 1970, when probably a half-million people — and possibly a million — were drowned. Surging waters, as much as fifteen feet higher than normal, wiped out all life on some islands and destroyed 1,100,000 acres of rice paddies in the East Pakistan delta. Starvation followed and then revolution, with the establishment of Bangladesh — a country with a population of 75 million, many of them vulnerable to future catastrophe.

Europe does not have hurricanes, but wind and waters, especially in the North Sea, can combine with devastating results. Time after time, gales blowing in a constant direction and at constant speed have generated storm surges that have swept over the shores of England and the Low Countries. The surges are known as sea bears, from Low German *see booren.*

The twin storms of December 22 and 23, 1954, demonstrated the part played by resonance in such surges. They struck at an interval of thirty-six hours, which is the North Sea's period of oscillation. The second storm reinforced the surge from the first.

Harlingen, on the northern shore of Holland, experienced the highest water level ever recorded there, more than twelve feet above normal. The water in such a case slopes upward—perhaps by only one-third of an inch in two miles—over a great distance in the North Sea, and this culminates in a very high tide that inundates the shore.

There is a long history of such surges. One in the year 1099 is said to have killed 100,000 people in Holland and England. In 1287 a sea wall collapsed and the Zuider Zee was flooded, with a loss of 50,000. The same storm took 500 lives in England. In 1421 the sea broke through Zuider Zee dikes and flooded seventy-two villages, with a death toll of 10,000. There were other great surges on the Dutch coast in 1570, 1825, 1894, and 1916.

On January 31, 1953, a terrific storm combined with a spring tide to breach dikes in a hundred places, flooding 4 million acres (one-sixth of Holland) and drowning 2000 people. On the British coast 300 hundred were killed.

For centuries the Dutch have been fighting such storms, building dikes, dams, and canals, pumping out lowlands with windmills to claim and reclaim land. After the 1933 storm the Netherlands embarked on a $650 million program to build twenty-five miles of new dikes. Doubtless the battle will go on as long as the winds blow on the North Sea. The land has been sinking and the ocean continues to rise, which means more, rather than less, distress.

Venice has a similar problem, on a much smaller scale but particularly tragic because the city is one of the world's greatest treasures.

According to legend, Venice was settled in A.D. 421 by fugitives from Attila the Hun, who found refuge on a group of islands. The logic of hindsight says that although they saved their skins they picked a poor place for a city, but this fails in appreciation of the magnificence that Venice achieved. Rather, it should be said that the city's site and canals, which brought it Byzantine splendor and Renaissance glory, ill equipped it to deal with modern times, tides, and industrial pollution.

The Adriatic Sea at Venice has a mean tide range of only 1.7

feet. A few centuries ago the tidal flow was enough to keep the Lagoon and canals open for vessels bringing back riches and to flush away wastes (Venice has never had a sewerage system). The situation has changed.

The city is built upon piles sunk into the mud and it stands only inches above high tide. Like the seas everywhere the Adriatic has been rising, and at the same time Venice has sunk, in some places as much as 7 inches in a half-century. This settling increased in recent years, thanks to the construction of a modern industrial center, Porto Marghera, on the mainland behind the city. Channels for big oil tankers were dredged and enormous amounts of fresh water were drawn from the subsoil for industrial purposes: about 7000 wells were drilled. Lowering of the water table caused greater subsidence and, in the opinion of some, the deeper channels caused tidal currents to run faster and unpredictably.

From time to time through the centuries there have been tidal floods. Many people were drowned when water invaded homes in 1250. Sea walls of wicker and clay were build in the fourteenth century and a stone sea wall was started in 1744, with later improvements and periods of neglect. The *acque alte*—high waters—occurred fifty-four times between 1866 and 1966, the majority of them in the final decade. In the period of 1972–74 the water rose high enough to cover pavements 295 times.

On November 3, 1966, all of northern Italy was struck by a sirocco, that fierce wind from the Sahara Desert which picks up moisture from the Mediterranean. High tide at Venice that night was higher than usual and, to the consternation of citizens, no ebb came. High tide the next morning drove the waters still higher; and the next ebb began four hours late, after the wind finally dropped.

> Across from the city, on the long narrow strip of land that protects the Lagoon from the Adriatic, events so unprecedented were occurring that if they had been known of in Venice everyone might well have fled in panic.

So wrote Stephen Fay and Phillip Knightly in *The Death of Venice*.

> At Pellestrina huge waves pounded the massive murazzi, the stone sea walls built in the late eighteenth century to keep the Adriatic at bay, smashing free huge blocks of marble and flinging them aside like pebbles, until the walls cracked and then collapsed, and the water rushed in. On wider stretches of the peninsula the sea rose and covered the land until the Adriatic and the Lagoon became one.

The flood reached 6½ feet above normal high tide, waves broke on the arches of the Ducal Palace, 5000 people were homeless, and damage totaled $60 billion.

At the same time, there were floods throughout southern Europe, with nearly 150 people killed. The sirocco caused flooding of the Arno River at Florence, with immense damage to works of art and other treasures, and of course this had nothing to do with the tides. A great program of restoration was undertaken.

Circumstances at Venice were more calamitous, in a way, because the problems are more complex; to some, they seemed almost beyond solution. The city has the afflictions of other urban centers—such as a polluted atmosphere eating away great sculpture, plus its own unique problems: The tides had to be allowed to flow in for sewage disposal, and industry insisted on the continued passage of ships. Foundations of some of the great old buildings have weakened, resulting in structural damage; and high waters have loosened plaster walls. Some 80,000 of the 180,000 residents (as of 1945) moved to the mainland. The issue arose of industrial jobs for workers vs. maintaining Venice as "a playground for the rich." Italian politics impeded progress in making improvements for which millions of dollars were raised all over the world. Fay and Knightly concluded: "Venice is dying and there is no hope of saving her."

This seems to have been too pessimistic an opinion. Remedies are being found. Many of the wells were capped and an aqueduct was built to bring in water from the mountains. Water pressure is building up again in the subsoil, and there is hope that the city will rise about an inch in the next two decades. Plans for sewage disposal were developed. Heating of homes has been switched from oil to nonpolluting methane or electricity. Construction of another industrial center was postponed. Plans were made for piping oil to factories rather than delivering it in tankers. Extensive research has been conducted on methods of installing barriers, perhaps temporary, at times of the *acque alte*.

One thing is certain: the tides cannot be reorganized. The Adriatic has a special affinity with siroccos. About 500 miles long, it has a period of 21.4 hours and is responsive to tidal forces of the Ionian Sea. It is divided into two basins, the lower one having a depth of as much as 4100 feet, creating a funnel for waters moving into the upper one, which has a maximum depth of only 790 feet. There is an amphidromic movement in the upper basin. Tides at Brindisi, on the heel of the Italian boot near the mouth of the Adriatic, are slight but when a sirocco pumps them they can be substantial at Trieste and Venice, at the head of the sea.

Venice will survive, as the world will survive hurricanes, typhoons, cyclones, and destructive nor'easters. But millions of people who have the bad luck or poor judgment to live within reach of storm tides may not survive. Looking backward may help us live now and in the future more defensively.

If one wants cause for useless worry, it can be provided by mention of the highest possible tide that can be provided by astronomical conditions. This occurs when the moon and sun are simultaneously at proxigee–syzygy, at the same declination, and with the earth at its closest annual approach to the sun—a condition that occurs about every 1800 years, in the opinion of some scientists. Various dates have been given for this event—3500 B.C., 1900 B.C., 250 B.C., and A.D. 1340, with the next one due in the year 3300, when, happily, none of us will be present.

NOTES

The introductory quotation is the terse statement made by Paul Godfrey, professor of ecology at the University of Massachusetts, after he had observed the devastation at Hull, Mass., caused by the "Blizzard of 1978." He has been conducting a study of coastal barrier islands for President Carter. The statement appeared in *The Patriot Ledger*, Quincy, Mass., February 27, 1978.

It will be apparent to the reader of this chapter that the principal source of information was Fergus J. Wood. He kindly supplied me with advance proofs of his book, *The Strategic Role of Perigean Spring Tides in Nautical History and North American Coastal Flooding, 1635–1976* (U.S. Government Printing Office, Washington, 1978) and with a copy of the supplement to it prepared after the 1978 storms. But beyond that, he provided the kind of patient, stimulating, professorial tutoring to the writer that any faltering student is lucky to receive.

Information about the 1978 storms was drawn chiefly from the excellent coverage by *The Boston Herald American, The Boston Globe, The Patriot Ledger*, the *Hingham Mirror, The Hingham Journal*, the National Weather Service, the National Ocean Survey, and personal experience. *Northeast Blizzard of '78, Natural Disaster Survey Report, 78-1*, a report of the disaster team to the administrator of NOAA, was released in November, 1978.

Governor William Bradford is quoted from *History of Plimoth Plantation* (Secretary of the Commonwealth of Massachusetts, 1898). A valuable source of information about all catastrophes is *The Great International Disaster Book* by James Cornell of the Smithsonian Astrophysical Observatory (Charles Scribner's Sons, New York, 1976), who also gave personal assistance. The story of the 1938 hurricane is told in *A Wind to Shake the World* by Everett S. Allen (Little, Brown, Boston, 1976). Albert Defant's *Ebb and Flow*, previously cited, provides analyses of North Sea storms.

The Death of Venice by Stephen Fay and Phillip Knightly (Praeger, New York, 1976) reviews the Venetian problems. Excellent articles on Venice were written by Dora Jane Hamblin (*Smithsonian*, November 1977, Vol. 8, No. 8, p. 40-53) and Sari Gilbert (*The Washington Post*, September 22, 1975).

An informative article on coastal storms, "Swept Away," by Ben Funk, appeared in *The New York Times Magazine*, September 18, 1977. An article, "Coastal Flooding" by Cynthia J. Brodnax, chiefly on Fergus Wood's studies, was published in *Sealift* (Vol. XXV, No. 7, July 1975) a publication of the Navy's Military Sealift Command). The New England Marine Resources Information Program (NEMRIP) *Information* bulletin No. 75 (University of Rhode Island, Narragansett, R.I.), provides a concise summary of the January 8-12 and February 6-9, 1974 tidal floodings, as does *Science Year* (The World Book Science Annual) 1975, pp. 328-9. The increased potential for these flooding events was predicted in the Science sections of *Time* and *Newsweek* for January 7, 1974.

Luigi Amaduzzi of the Embassy of Italy and Dr. Marco Miele, director of the Italian Cultural Institute in New York, provided additional information.

NOAA's Ark

National Ocean Survey rides in the vanguard of
agencies concerned with the difficult problems
of land use and environment, boundaries and
coastal resources, crowded shipping lanes and
crowded skies, and the confrontations between
man and ocean along the water's edges. IN-
TRODUCTION TO NOAA'S NATIONAL OCEAN
SURVEY

A DOZEN MILES northwest
of Washington, D.C., is one of those centers of federal
decentralization—the Rockville headquarters of the National
Oceanic and Atmospheric Administration. The cluster of tall
buildings, handsomely landscaped and unostentatiously labeled,
is set apart in a Maryland valley where raw gashes in the red clay
reveal the recentness of construction.

NOAA (pronounced Noah) is the 1970 product of reorganization, under the Department of Commerce, of various agencies including what had once been the U.S. Weather Bureau and the
Coast and Geodetic Survey. It is now composed of the National
Weather Service, National Ocean Survey, National Environmental Satellite Service, Environmental Research Laboratories, Environmental Data Service, and National Marine Fisheries Service.

Perhaps the earliest published tide table in England was that issued by St. Albans Abbey in A.D. 1250, listing the times of high water at London Bridge—inaccurately when judged by modern standards. The Catalan Atlas of 1375 gave the time of high tides on both the Briton and Breton shores of the Channel. Through the next four and a half centuries information about tides was gradually accumulated, but even after Newton expounded the tidal theory, data on actual tides was often faulty. Checking on water height might be the duty of an unreliable wharf watchman. Almanacs were commercial ventures and their method of calculating tides was a trade secret.

The first great authority on navigation was an American, Nathaniel Bowditch, whose name is almost holy to every man who has served on the bridge of a ship. Bowditch was born in Salem, Massachusetts, in 1773, the son of Habakkuk Bowditch, who was fourth in a line of shipmasters. The father was an improvident, heavy drinker, and at the age of ten Nathaniel was put to work for a cooper; later he was apprenticed to a ship chandler. With borrowed books he educated himself, displaying early a genius for mathematics. He went to sea as a clerk, learned everything he could about navigation, and discovered that nautical charts and tables for the East Indies were extremely untrustworthy.

A Newburyport publisher wanted to bring out an American edition of *The Practical Navigator,* a standard manual by John Hamilton Moore of England; he asked Bowditch to revise it. Bowditch found 8000 errors and prepared new tables for the edition, which was published in 1799. Then Bowditch wrote a completely new text, published in 1802 as *The New American Practical Navigator.* It was not only sold in this country but distributed in England by Moore's publisher.

Bowditch had developed a new method of determining a ship's position by observing the angular relation of the moon to a planet or a star. He was primarily concerned with navigation but he devoted a chapter to the tides, with rules and tables for finding the moon's age and therefore the time of tides any place in the world. He wrote:

> Many pilots reckon the time of high-water by
> the point of the compass the moon is upon at
> that time, allowing 45 minutes for each point.
> Thus on the full and change [new moon] days,
> if it is high-water at noon, they say a S. by E. or
> N. by W. moon makes a full sea; and in like
> manner for any other time. But it is a very inac-
> curate way of finding the time of full sea by the
> bearing of the moon, except in places where it is
> high-water about noon on the full and change
> days.

Bowditch offered a table with basic tide data for the principal American ports and such farflung places as the Amazon River and Cape Blanco in "Negroland."

In a biography, *Yankee Stargazer,* Robert Elton Perry wrote:

> It was due to Bowditch more than any other
> man that Yankee shipmasters came to be ad-
> mired for their ingenuity as navigators. When
> Bowditch came along, Yankee shipmasters were
> already known in most ports of the world for
> their smart seamanship. . . . The principal
> weakness of the Yankee shipmaster was his navi-
> gation. Bowditch was a major influence in
> changing this weakness into a strong point. . . .
> He gave the American seamen a book that
> could be understood and trusted.

Bowditch's book went through thirty-two editions by the time of the Civil War, and since 1868 its publisher has been the federal government. Under the title *American Practical Navigator* (Bowditch) it is now issued by the Defense Agency Hydrographic Center in two volumes. Altogether, it is estimated that 930,000 copies of the manual have been sold in its seventy editions.

Bowditch made his last voyage as a master and part-owner of a ship loaded with pepper and coffee from the East, bringing her safely into Salem Harbor on Christmas Day, 1803, in a blinding snowstorm while other ships prudently waited outside for the weather to clear. He never stopped studying. He translated four abstruse volumes of Laplace's *Mécanique céleste*, became presi-dent of an insurance company, president of the American

Academy of Arts and Sciences, and a member of the Harvard University Corporation.

The U.S. government established a "Survey of the Coast" in 1807 and published tide predictions for the first time in *The American Almanac* in 1830: one-a-day high water predictions for Boston, New York, and Charleston, with time differences for 96 other stations.

In that same year the *British Almanac* was first published, with tide predictions prepared by Sir William Lubbock, an English astronomer. Lubbock had discovered how erroneous the privately published tables were, and he and Joseph Dessiou began collecting accurate data. He interested his former tutor at Cambridge University, William Whewell, in the enterprise, and arrangements were made to measure the tides during a fortnight in 1835 in 500 places in Great Britain and 28 others on both sides of the Atlantic. Drawing cotidal lines, Whewell discovered that there were rotating tides in the North Sea. For many years afterward, there was controversy as to whether rotating tides existed in other seas. In fact, nearly three-quarters of a century passed before the argument was settled by Rollin Harris's studies of amphidromes.

Through the early days all the mathematical work for tide predictions had to be done by hand, of course — a laborious process. In 1867, Sir William Thomson, later made Lord Kelvin, devised a method of using harmonic analysis; five years later he developed a machine to do the work. William Ferrel, a mathematician for the U.S. Coast and Geodetic Survey, also developed a machine that was used in making tide tables from 1885 to 1914. That brings us back to the Harris–Fischer machine and to NOAA.

The tide-predicting machines were essentially analogue computers. The turn of a gear — speaking in very rough terms — was the mechanical analogue of the turning of the earth or the revolving of the moon. One could describe the various tidal forces mathematically, adjust the machine to represent them, turn the crank and, through harmonic analysis, come up with the sum of the effects.

The astronomical factors in determining tides are known as harmonic constituents. The government manual lists about 150 of them, but the number is really infinite since a change in one constituent affects others and thus a new constituent must be added. The most important ones are:

M_2　Main lunar (semidiurnal).

S_2　Main solar (semidiurnal).

K_2　Soli-lunar (semidiurnal) due to changes in declination of sun and moon.

N_2　Larger elliptic (semidiurnal).

O_1　Main lunar (diurnal).

K_1　Soli-lunar declinational (diurnal).

Only nineteen constituents were used in predicting tides between 1885 and 1911 (for 1912 tables). Then eighteen more were introduced, though twenty-five rather than thirty-seven are usually employed at present in preparing tables. In special cases, more may be used. When oil was discovered in the Cook Inlet in Alaska, that long, narrow arm leading up to Anchorage, 114 constituents went into tide calculations to make sure no unforeseen problems would be encountered by tankers.

Particular attention was also given to prediction of tides in waters leading to Valdez, the port at which tankers pick up oil from the trans-Alaska pipeline. Just outside the harbor are the Valdez Narrows, a four-mile channel which is only 1200 yards wide and has the dangerous Middle Rock as a hazard. Before any oil was transported, a tanker loaded with sea water made test runs. Predictions for Valdez were included in tables for the first time in 1979.

Use of a large number of constituents is made possible by modern computing methods. In 1966 the government abandoned the Harris–Fischer machine and began processing tidal data on a digital computer. The old machine could prepare the table for one year at one station in one man-day. The computer does the same job in thirty seconds.

Not only is the computer enormously faster but, as demonstrated in the case of the Cook Inlet, it has a tremendous capacity for complex data. This brings up a touchy point involving the National Ocean Survey and the National Weather Service, which coexist in NOAA's ark in tremulous bureaucratic equilibrium. NOS predicts the tides and NWS predicts the weather. In anticipating storm surges, NWS looks at the NOS tide tables and estimates how much the tides will be increased by winds.

In his book on perigean spring tides (discussed in Chapter 5), Fergus Wood developed a method by which "killer tides" could be predicted and evaluated. The degree of hazard depends, of course, on the presence of onshore winds which can drive high tides higher, and weather conditions cannot be predicted very far in advance. But by using some additional tidal constituents in the tide calculations, the potential for coastal flooding could be more precisely determined. Wood wrote:

> The nearly phenomenal advances during recent years in high-speed electronic data processing and related technology have rendered possible vastly more complex computational procedures. At the same time, new requirements for even greater precision in certain local tidal predictions have arisen. These have come about as the result of the burgeoning development of coastal communities, the increasing establishment of vacation homesites, beach cottages, and condominiums ever closer to the high waterline, and the expanded and proliferating recreational use of the coastal zone.
>
> Such growing demands make a review of the previous tide-computing methodology both necessary and desirable. . . . Salient investigations should include special attention to the delineation of factors useful in analyzing, predicting, and gauging the probable intensity of tidal flooding well in advance of catastrophic events.

As was explained in the last chapter, coastal flooding can be worsened at the time of certain astronomical circumstances not only by an increased height of the tide but also by the increased speed with which a tide rises and by the increased duration of a high tide. Perhaps NOS and NWS will find ways to take into account the complexities described by Fergus Wood but, at this writing, they have shown no inclination to do so. NOAA does not plan to issue special warnings of high-risk tides as it did for those of January 8 and February 7, 1974. NWS bureaus will continue to use NOS tables and their knowledge of local weather and tidal conditions in predicting storm surges despite a failure to predict accurately, among numerous instances, the four tidal floodings experienced in 1978 on both the East Coast and the West Coast during the perigean spring tides and pseudo-perigean spring tides of January 8–9 and February 6–7 (see Chapter 5).

Tide tables are prepared under the direction of Donald C. Simpson, a prematurely white-haired State-of-Mainer, laconic and apparently unflappable, who is chief of the Predictions Branch of the NOS Oceanographic Division. He and his staff of nine are responsible for the accuracy of a very large volume of data. It takes some 1460 items of data to describe the tides at just one station for one year. Tide predictions are completed and committed to magnetic tape about two years in advance, go to press about a year later, and are ready for distribution six months or so before the year to which they apply.

Before the tables are printed there must be a great deal of checking and rechecking and computer programming. Jack E. Fancher, red-bearded chief of the data-processing section, supervises a score of people in preparing raw data for harmonic analysis and feeding it by wire to an IBM 360 Model 65 computer in Georgetown. If there is a snag, problem data can be recalled for visual examination on a scope, so that a correction can be made.

NOS publishes four volumes of tide tables each year—one for the east coast of North and South America, including Greenland; one for the west coast of North and South America, including the Hawaiian Islands; one for Europe and the west coast of Africa, including the Mediterranean Sea, and one for the Central and Western Pacific Ocean and the Indian Ocean.

I have been discussing the processing of tide data as if nothing but mathematical abstractions were fed into the computer. This is not the case. The numbers on which predictions are based are actual readings from a network of tide stations, a total of 150 including those in Hawaii and Guam. In addition, special studies are made in specific locations by six-man crews which install temporary tide gauges. New Jersey, South Carolina, Florida, and California waters are among those which lately have been given concentrated attention. At any one time, NOS is receiving 500 or more sets of tide readings. It also trades data with foreign countries, notably Great Britain for, although the Empire has shrunk, the British publish comprehensive tables for the world.

A tide gauge operates on the same principle as the floating copper ball in a toilet tank. When the tide rises, a brass float (shielded from waves in a pipe-like well) is lifted, and the movement is transmitted by a wire running over a pulley in a recording device. The time-honored mechanism had a revolving drum with a sheet of graph paper on which a pencil recorded changes in water level. This machine is being phased out and replaced by an ADR (for Analogue to Digital Recorder) which punches tape every six minutes. A tape runs for a month and is then sent to NOS, where a machine automatically reads the data and transfers them to magnetic tape.

The permanent stations are generally located on wharves and are housed in small wooden "shacks" to protect them from the weather. Although they operate automatically, each one is tended by a local observer whose task it is (for a fee of fifty dollars a month) to visit it five times a week. Each station has a "staff," a graduated measuring board fastened to the wharf. The observer checks on whether the water level on the staff is the same as that being registered by the machine. If the well is clogged with seaweed (or rubbish) he cleans it out and during the winter he may have to pour kerosene into it to prevent freezing. If the recorder is an old-fashioned one, he winds the clock which operates it; if it's electrified he changes batteries from time to time. He mails a written report each week and once a month puts on a new roll of tape and ships the old one to Rockville.

Solar power is being tested for operating the recording mechanisms but no substitute has been found for a human being who can make adjustments and telephone headquarters when the weather wipes out a station. During the harsh winter of 1977, ice destroyed fourteen stations on the Atlantic coast and damaged nineteen others. Four California stations were lost during the storm of January 1978.

One new kind of gauge uses gas pressure in sensing tidal changes. It can be rigged for telemetering data, which may make stations fully automatic in the future. Such is the gauge at Washington, D.C., situated on the Wilson Boat Lines pier on the Potomac River. It telemeters data to NOS at Rockville, where the tidal curve is shown visually on a scope. But even that station must have an observer — in this case a bus driver who makes a visit to the shack when he stops his bus at the pier.

More typical is the Boston observer, Fred W. Wiley. His grandfather was a seaman, his father was in the Coast Guard, and he has long been on intimate terms with the ocean. Wiley joined the Navy during World War II, and after the war he went into the Coast Guard as an electronics technician. He spent most of his time calibrating instruments for weather ships. Later he transferred to the Weather Service and continued to do the same sort of work. He occasionally went to the tide station to adjust instruments. When a new observer was needed, he took over the job as an adjunct to his retirement.

Wiley lives nineteen miles from Boston, so fifty dollars doesn't much more than pay for transportation. He arrives about noon each weekday at the shack, a tiny shingled building on a dock at the entrance to the Fort Point Channel, and checks the clock with a radio time signal. He dips some water out of the channel and determines the specific gravity as a way of measuring salinity and makes a record of water and air temperatures and of the weather. Tide readings are sent automatically by wire to the Weather Service office at Logan Airport, which he can see across the harbor. After completing his duties he waters his potted plants, which give the shack the look of a cozy little cottage.

Fred Wiley is a robust 58-year-old but when he does get sick he sends a son to the station. During the Blizzard of '78 a coast guardsman took over. Such dependability is notable too at the

Fred W. Wiley, Boston tide observer for the National Ocean Survey, checks the tape machine in the "shack" at the entrance to Fort Point Channel. *(Photo by author)*

station on Cedar Key, Florida, twelve miles south of the mouth of the Suwannee River on the Gulf of Mexico. Nearly a half-century ago, W. Randolph Hodges, owner of one of the first fish companies on the key, became observer. Then the job passed to his son, W. Randolph Hodges, who has also served as state senator and director of the Department of Natural Resources. In the third generation, Gene Hodges, now a state representative, served for a time and the present observer is his brother, 29-year-old Hal Hodges, proprietor of a hardware, bait, and tackle store.

The Cedar Key shack is on the city dock and Hodges usually visits it late in the afternoon. He likes to keep track of the tides because he is a mullet fisherman; although the mean range there is only 2½ feet, the incoming tide is the best for fishing. Crab fishermen are most interested in finding out from him the temperature of the water.

For tide data to have much meaning to NOS scientists at Rockville, they must know the datums for the stations. (Datum, in this case, is the technical term for the base plane to which the rise and fall of waters is related. The plural takes an "s.") The principal ones are *mean sea level; mean low water,* which is used on the Atlantic Coast; and *mean lower low water,* used on the Pacific coast, where the tides are generally mixed.

Roughly speaking, mean sea level is the mean height of the water at a particular location, derived mathematically from hourly measurements over a nineteen-year period.

Sea level, of course, is relative to the shore. Piers on which tide stations are located can rot and sink into the mud, so the top of the tide staff is leveled with a permanent mark — a bench mark set in concrete or solid rock nearby. In fact, a requirement is that there be at least three bench marks within a mile of a station. A bench mark is a bronze disk, $3\frac{5}{8}$ inches in diameter, carefully placed with the aid of surveying instruments. It is used for mapping as well as reference for tidal observations. From time to time the level of a bench mark must be checked with that of the tide staff and the tide gauge.

After learning in school that Pikes Peak is 14,110 feet above sea level and Death Valley is 282 feet below sea level, one gets the impression that sea level is a sacred sort of constant like the speed of light or the date of Christmas. In the world of tides, that is by no means the case. Sea level, though measurable, is variable.

In the first place, there has been the long-term change since the Ice Age when so much water was locked up in glaciers that sea level was as much as 520 feet lower than it is today. As the icecaps continue to melt, sea level continues to rise. Scientists do not agree as to the rate. There is a place on the Connecticut shore, west of the mouth of the Connecticut River, which offers evidence that sea level has increased thirty-three feet in the last 7000 years.

Rainfall, winds, temperature, salinity, and atmospheric pressure variations are responsible for daily and seasonal changes in sea level, which may amount to a half-foot or more. Along the Atlantic coast, sea level is lowest during the winter months and highest in late summer and fall. On the Pacific coast, from San Francisco to San Diego, sea level is lowest in the springtime and it peaks in September. At Seattle it is high in mid-winter and low throughout the summer.

In some places, rising of the sea level, as determined by tide gauges, actually means that the land is sinking. Or if the sea level is falling, it may mean that the land is rising. Ever since the Ice Age, lands which were covered with glaciers have been rising — relieved of the weight of mile-thick ice. Elsewhere, tectonic changes, deep below the surface, may cause the land to rise or fall.

Careful analysis of tide records is required to determine just what is happening. This is important to understanding changes in the earth and may be very important in a practical way. The most notable example is on the Gulf shore of Louisiana and Texas, where there has been a rapid rise in the sea level since 1940 — as much as a foot and a half at Eugene Island, Louisiana. Such a rise is attributed to a sharp, and even alarming, subsidence of the land, believed to be due to the pumping of oil and groundwater from beneath the surface. Measurements in Louisiana go back only to 1940 but the trend there has been nearly four inches a year. At Galveston, where measurements started in 1908, the trend is about two-thirds as great.

"We are keeping a very close watch on that area," says James Hubbard, chief of the NOS branch that oversees datums. "There has been extensive residential construction along the shore, especially in certain parts of Texas. With land subsiding as much as it is, the vulnerability to hurricane flooding is increasing. The oldest tide station in the country, established at San Francisco in 1855, has shown the sea level there going up steadily but there has been no sharp rise as on the Gulf coast."

Hubbard, a lean, thoughtful man, explains that the branch's main function is to compute datums for use in hydrographic charts. Tide readings collected now are compared to those in the last nineteen-year cycle, from 1941 to 1959. If they agree, there is no problem. If they do not, adjustments must be made in the datums. The basic standard is that of mean high water in 1929, a date selected arbitrarily when the widely varying figures from many locations were made as compatible as possible.

The datum for *mean high water* is particularly critical because in most coastal areas the high-water lines serve as boundaries for states and private property. If oil is discovered on a shore, the precise location of mean high water can make a difference of millions of dollars to property owners.

Hubbard and his staff of ten spend a large amount of their time in answering inquiries about boundary lines and appearing as expert witnesses in disputes between states over boundaries and in suits over property rights. They were even called upon for evidence in a murder case (where the height of the tide was a factor) and were asked by a movie company about the best tide for pushing an automobile into the sea.

NOAA's publications are the authoritative sources of information about tides and currents but there are many commercial publications that offer the data in more or less complete form (from NOS tables) along with information about everything from astrology to knots. *The Old Farmer's Almanac,* established in 1792, gives sketchy tide references; and although one might suppose a farmer in the Midwest wouldn't be interested, the editor claims that inland people check the tide hours for the best times to go fishing in tideless fresh water.

weak currents." This does not always hold true. In the Race, at the eastern end of Long Island Sound, the tidal range is 2½ feet and the current runs as high as five knots. At the western end of the Sound, the range is 7½ feet and the current is less than one knot.

Tides and tidal currents in Long Island Sound and the waters around New York City are extremely complicated. H. A. Marmer of the Coast and Geodetic Survey, who studied them carefully, wrote of the Sound:

> The range of the tide increases in a fairly uniform manner from the eastern end despite the fact that at its middle the Sound has a greater width than at the ocean entrance. . . .
>
> Another peculiar feature of the tide in Long Island Sound is that there is but little difference in time of high water or low water from one end to the other; that is, over the greater part of its area the tide is nearly simultaneous. . . .
>
> Over the greater part of the Sound the current is very nearly simultaneous. Furthermore, in relation to the tide it is found that the strength of the current comes approximately midway between the times of high and low water, while the slack of the current occurs about the times of tide.
>
> In Long Island Sound, therefore, the tidal movement presents none of the characteristics of progressive-wave motion. . . . It becomes evident at once that the tidal movement in Long Island Sound is of a stationary-wave type, exemplified by the half of a simple stationary wave in a tank of water.

Making calculations on the basis of the eighty-mile length of the Sound and an average depth of sixty-five feet, Marmer found its period of resonance to be 11.8 hours—close to the 12-hour period of the ocean tide entering from the east and therefore having the resonance needed for a stationary wave.

Tidal waters flow into and out of Long Island Sound not only at the eastern, ocean end but also at the west end, setting up swift

106

currents of 2½ knots in the East River and 5 knots or more at Hell Gate in the Harlem River. It takes the tide only one hour to move twelve miles up the Hudson to Spuyten Duyvil but three and a quarter hours to go fourteen miles up the East River to Throgs Neck. The schedule of the tide in Long Island Sound is still different. The result is that there are pronounced differentials in levels of water at certain times. And, since water runs downhill, strong currents are the result. Each day, Long Island Sound pours 300 million cubic feet more water into New York Harbor through these channels than it receives from them on flood tides.

Peculiarities of tidal patterns east of Long Island can help or hinder yachtsmen, Eldridge points out. The ebb current runs toward the east in Long Island Sound but toward the west in Nantucket and Vineyard Sounds and Buzzards Bay. Flood currents are in the opposite direction. A boat bound east from Long Island Sound can have a fair current on the ebb through the Race and then ride on the flood current toward Newport and Cape Cod. Sailing toward the west it can also use such advantageous currents.

I have been writing of currents as if they always moved in two directions as the tide swings back and forth, and so they do, usually, when enclosed by shores. But in the open ocean, and some other places, they can be rotary, swinging through all points of the compass during a tidal cycle. This occurs in the case of amphidromes; it can be due to the shape of coasts or bottoms or it can be caused by changes in the direction of gravitational forces or by the confluence of tides from different directions.

Nantucket Shoals, southeast of Nantucket Island, has a pronounced rotary tide. (Incidentally, the shoals have glacial moraines as their base, but they consist mainly of waves of fine and medium-grained sand and silt deposited in the ebb and flow from the Gulf of Maine.) Marmer reported that the current circled clockwise, from a northeasterly direction at the time of high tide in Boston through the compass and back to northeast, with a maximum velocity of 1.3 knots and a minimum of 0.8. Speed and direction change with phases of the moon but the currents still maintain their rotary motion.

There are rotary currents off the Pacific Coast, too, and these, as might be expected, reflect the diurnal inequality of the tides. A rotary about ten miles from San Francisco Bay makes a small rotation (that is, with weak currents) through the higher high waters and then a large one (with strong currents) through lower low waters into lower high waters.

The wind is the principal driver of the great ocean currents, such as the Gulf Stream, on which the tides have no influence. The wind is the ruler of sailing craft, too, but if it relaxes in its rule, currents take charge. As a novice sailor I learned that lesson while heading for a mooring at dusk against an ebb current and with a gentle breeze over the bow. The channel was narrow and this meant that short tacks were unavoidable. Although I could gain a few feet on a tack, the current promptly swept the boat back again when I came about. Not until slack tide and nightfall could I reach the harbor.

Tidal currents can be crucial in sailing races and there is a premium on good judgment in using them. It has been said that by finding a current with a half-knot advantage, one can gain 500 feet over competitors in ten minutes. Racing tactics can be complicated and I shall not go into detail. One generalization is that the current is weaker along shores except where it is deflected by points of land. So the skipper who is familiar with rocks and shoals at various tides can hug the shore when bucking a tide and make greater headway than boats in deep water where the current is stronger. Of course the reverse is true if moving with the current.

There are times when one cannot avoid an adverse current. Robert N. Bavier, Jr., advises:

> An anchor is a valuable asset throughout a light wind race whenever the tide is foul. If the breeze peters out you are apt to be losing ground even though you are still sailing through the water. Under such conditions, check your progress by taking ranges on landmarks. If you are losing, drop the anchor over quietly and, if possible, without your competitors seeing it. You may be

able to gain quite a lead before they realize they
are going backward.

Lee-bowing is a tactic that sometimes helps a close-hauled boat sailing against the tide. By pinching—sailing closer to the wind than one usually would—and keeping the set of the current on the lee bow to push the boat to windward, one can sometimes fetch a weather mark or round a point without tacking. But Bavier thinks the resultant loss of speed may often be costly.

Finding eddies in the tidal current can be helpful, according to a treatise by Harold Augustin Calahan and John B. Trevor, Jr. They wrote:

> On occasion we have sailed through a series of such eddies, some of which were whirling clockwise and others counter-clockwise, and by watching the motion of the water we struck eddy after eddy on its right side and were carried forward against a strong tide, with barely enough wind to give us steerage-way. In one race in particular we overtook and passed eight boats within a short period of time. Before we found the eddies we were nearly two miles behind the last boat, and after we sailed through the eddies we were fully a mile and a half ahead of the first boat.

But they warned:

> Do not attempt to sail through strong tide rips when a strong wind is making up against the current. The going is unbelievably bad, and is downright dangerous even for a well-found boat. The seas are high, short, breaking, and unpredictably crazy. They will break on board from every direction.

Most notorious of waters in which no boat is safe are the rivers in which there is a tidal bore—the steep wave with which the tide arrives in a few rivers of the world. Victor Hugo's newly married daughter and her husband, Charles Vacquier, were drowned in 1843 in front of Hugo's home when the bore in the Seine caught their small sailboat.

Tidal bore in the Petitcodiac River at Moncton, New Brunswick, where high tides from the Bay of Fundy drive water back up the river. The bore has become smaller in recent years. *(New Brunswick Department of Tourism)*

The Seine bore is known as the *mascaret* and it reaches a frightening height of twenty-four feet — or did before dredging of the river caused some reduction. It is highest when a west wind drives the tide, but is pronounced at the time of spring tides. It advances at fifteen miles an hour — very fast indeed for a raging wall of water.

Bores occur in streams with funnel-shaped shores and shoaling bottoms where tidal ranges are high. The incoming tide wave is retarded by friction in the shallowing water until it moves more slowly than the current and builds up into a turbulent crest.

One of the most famous bores is that in the Tsientang Chiang, which empties into the China Sea at Haining, south of Shanghai. In 1888 a Captain Moore of H.M.S. *Rambler* was making a survey of the river when he had his first experience with the bore, which rushes in at eighteen knots with a wave as high as twenty feet. Though the vessels in his flotilla were anchored and had their engines running at full speed, they were swept three miles up the river.

Having had long experience with the bore, the Chinese used it to their advantage, the captain reported. They would beach their junks on a protected shelf at high water, where they would be sheltered from the violence of the bore, and then on the next tide they would allow themselves to be carried upstream in its "after-rush."

According to Chinese legend, the bore represented the revenge of a general who had been assassinated by an emperor in ancient times. Its violence was lessened by building a pagoda on an embankment which the bore had damaged and by holding an annual ceremony of throwing offerings into the water.

A bore known as the *pororoca*, as high as fifteen feet, occurs in the Araguary, Guama, and Guajara rivers at the mouth of the Amazon. The turbulence there as the Amazon, which has the greatest flow of any river in the world, meets the ocean is said to be the reason it has not built a delta. The Amazon's muddy plume can be seen in the Atlantic twenty miles from its mouth.

The most famous bore in England is that of the Severn River, where a forty-foot tide occurs two or three days after new moon or full moon, producing a wave as high as five feet which travels thirteen miles an hour. A bore in the Trent is known as the *eagre* (other spellings are eager, aegir, and aiger). It can be more than six feet high and may have been the model for the one on the Floss River, in *The Mill on the Floss* (but ill-fated Tom and Maggie were apparently drowned in a storm surge rather than a bore).

The Orne and Gironde in France have bores. In Turnagain Arm and Knik Arm, near Anchorage, Alaska, at the head of Cook Inlet there are bores that reach six feet, just after low water, when conditions are right.

With record high tides in the Bay of Fundy one would expect a bore. For many years a great tourist attraction was the wall of foaming water, sometimes five feet high, which rushed up the Petitcodiac River at Moncton, New Brunswick, about two and a half hours before high tide. Construction to exclude tidal waters from the upper portion of the river reduced water velocities, and this resulted in rapid silting of the estuary. The bore today is a mere shadow of its former self.

Another New Brunswick spectacle, still to be seen, is the Reversing Falls at St. John, where Fundy tides of more than twenty-seven feet are not unusual. The St. John River passes through a narrow gorge, which impedes the incoming tide so that it is six to twelve feet higher than on the upstream side and forms a thundering waterfall. On the ebb tide, the river is higher above the gorge than the bay and there is a waterfall in the other direction. Only at slack, which lasts about twenty minutes, is the water calm enough for sailboats.

Samuel de Champlain had an encounter with the Reversing Falls on his first voyage in 1604 and recounted:

> The river is dangerous, if one does not observe carefully certain points and rocks on the two sides. It is narrow at the entrance, and then becomes broader. A certain point passed, it becomes narrower again, and forms a kind of fall between two large cliffs, where the water runs so rapidly that a piece of wood thrown in is drawn under and not seen again. But by waiting till high tide you can pass this fall easily.

112

Most famous of the ocean's turbulent perils was Charybdis, the whirlpool in the Strait of Messina, between Italy and Sicily, which Ulysses, in Homer's Odyssey, managed to avoid at the cost of having members of his crew gobbled up by Scylla, the monster who inhabited the opposite side of the Strait.

Homer undoubtedly had exaggerated tales from sailors to embellish for high literary effect. But there really is a whirlpool, caused by the six-knot current in the strait, with a treacherous rock named Scylla nearby; both are less awesome than in Homer.

Edgar Allan Poe let his imagination build upon legend for the celebrated story, "A Descent into the Maelstrom." This whirlpool off Norway, he said, was a mile in diameter, and he described it thus:

> The edge of the whirl was represented by a broad belt of gleaming spray; but no particle of this slipped into the mouth of the terrific funnel, whose interior, so far as the eye could fathom it, was a smooth, shining, and jet-black wall of water, inclined to the horizen at an angle of some forty-five degrees, speeding dizzily round and round with a swaying and sweltering motion, and sending forth to the winds an appalling voice, half shriek, half roar, such as not even the cataract of Niagara lifts up in its agony to Heaven.

Poe's protagonist in the tale, a Norwegian fisherman, had been drawn into the Maelstrom (Mosko-strom, he called it) on a seventy-ton schooner during a hurricane and survived by tying himself to an empty cask and jumping overboard. The ship was dashed to pieces on the rocks at the bottom but the cask, descending more slowly, returned to the surface at slack tide.

An old Norwegian belief was that water which whirled downward in the Maelstrom was carried through a tunnel all the way under the arm of Norway and Sweden, to emerge in the Gulf of Bothnia, at the head of the Baltic Sea. The facts are considerably less wild. According to F. G. Walton Smith, there are at least fifty powerful currents, with associated whirlpools, off the coast of Norway. In a narrow passage between two fjords at Bodo, speed of the Saltstrom reaches sixteen knots and "The noise of the flow, with its attendant whirlpools, is deafening."

Far more dangerous than the real maelstroms, and even more terrifying than imagined maelstroms, are the giant "rogue" waves of the open sea, which certainly reach 100 feet in height and, according to some calculations, may be even 198 feet high. Ship crews who have witnessed the very worst ones may not have survived to tell about it. These waves apparently occur when trains of waves happen to coincide and build up one monstrous freak. Tides are not involved, of course, but the superimposition of rhythms is of the same sort that can cause tidal phenomena.

NOTES

The epigraph is from "The First Relation of Jaques Carthier of St. Malo, 1534" which appears in *Early English and French Voyages, Chiefly from Hakluyt* (Barnes and Noble, New York, 1934).

The National Ocean Survey issues annually two volumes of *Tidal Current Tables*, one for the Atlantic coast of North America and one for the Pacific coasts of North America and Asia. It has various other publications, such as *Tidal Current Charts*, which are sets of twelve charts for specific waterways showing the direction and velocity of currents during each hour of the tidal cycle.

Marmer and Macmillan, already cited, were especially helpful on currents and bores. *Bores, Breakers, Waves and Wakes* by R. A. R. Tricker (American Elsevier, New York, 1969) is a comprehensive volume on those subjects. George Darwin, among others, describes the bore at Haining in *The Tides*, previously cited. On the Reversing Falls I have quoted from *Voyages of Samuel de Champlain, 1604–1618* (Barnes & Noble, New York, 1946).

I have quoted from *Wind and Tide in Yacht Racing* by Harold Augustin Calahan and John B. Trevor, Jr. (Harcourt, Brace, New York, 1936) and *Sailing to Win* by Robert N. Bavier, Jr. (Dodd, Mead, 1959).

"A Descent into the Maelstrom," by Edgar Allan Poe is quoted as it appears in *Great Sea Stories*, edited by Joseph L. French (Tudor Publishing Co., New York, 1950). F. G. Walton Smith sets the maelstrom to rights in *The Seas in Motion*, already cited.

VIII

Tides of Earth and Air

If the earth were of india-rubber the tides would be nothing, the rise and fall of the water relatively to the solid would be practically nil. LORD KELVIN

THUS FAR our attention has been given only to the tides in the ocean. But the pull of the moon and sun does not stop at the seashore. There are tides in the atmosphere and in the solid earth itself.

The *body tide*, as the tide in the earth is called, is quite different from the ocean tide. For one thing it responds to the vertical gravitational force of the moon rather than to horizontal tractive forces. And, instead of having a top layer lifted, as in the case of the ocean, the whole earth is deformed. It may not be a rubber ball but it bulges the way such a ball does when it is squeezed. Another difference, of course, is that there can be no currents in the earth, though compression of rocks takes place in the deformation.

Compared to the size of the earth, the bulge is very tiny and certainly not perceptible to us in going about our daily affairs. There is no place we can stand and watch the rise and fall, as we can watch the height of the sea change from the vantage point of a dock. Scientists have been trying for many years to measure the amplitude of the earth tide; their measurements, taken in different places, vary from 4½ to 40 inches. A bulge of either size, distributed over thousands of miles of the earth's curve, is very small.

There are two principal approaches to measuring such a bulge. The lunar pull should lessen the earth's gravitational force at a point under the moon. The standard unit for measuring gravity is the gal, named in honor of Galileo. The earth's gravity on this scale is 982.04 gals. The pull of the sun and moon on a ton of weight is only 0.2 of a gal. A gravimeter can be used to measure the lessening of the earth's gravity under tidal influence; this instrument is, in essence, a weight suspended on a spring.

The other approach is to measure the angle of the tilt in the earth's surface caused by tidal forces. This can be done with a horizontal pendulum which operates on the principle of a door that swings ajar when the door frame is not vertical. Or with a tiltmeter, which operates somewhat like a carpenter's level. In 1978 an M.I.T. student, Spahr Webb, made a series of measurements with two thirty-six-foot tiltmeters and found that the tidal tilt was about one one-hundredth of an inch.

A complication in making such measurements is that some tilting of the ground is caused by the burden of water on the edge of the continent, which changes with the ocean tides. High tides bring 100 billion tons of water into the Bay of Fundy twice a day and this weight has such an effect on the land that an attempt to measure the earth tide there would be useless: tilting by the earth tide would be hidden by distortion due to the ocean tide. Although Webb's measurements were made about thirty miles from the shore, he found that the tilting due to the direct earth tide and that due to the sea water burden were of approximately the same magnitude.

The effect of such ocean "loading" can be found even in data on earth tides obtained well inland. Strain meters, employing either laser beams or quartz rods, are used at observatories in New Jersey, Missouri, Colorado, Nevada, and California to measure distortion in solid rock. The studies showed that the tidal load must be taken into account even in the interior of the continent.

Rocket-hoisted satellites provide new ways of measuring earth and ocean tides. With laser, radar, and radio beams, very precise ranging can be done from high above the earth's surface. Such measurements are being made in a program called the GEOS-C Project, managed by NASA for the U.S. Departments of Defense and Commerce. Circling the earth at about 750 miles up, fourteen times a day, these satellites carry altimeters which, by sending out radar pulses, can measure tidal heights in both sea and land. Other measurements can be made by shooting laser beams from the ground at satellites and gauging the time it takes for the reflected light to return. Movements of the satellites, as their orbits are distorted by celestial and terrestrial gravitational forces, provide additional information.

In 1978, Seasat-A, first in a series of new satellites with which a Global Ocean Monitoring System is to be established, was launched. Circling the earth in 100 minutes in a near-polar orbit, it will measure the height of the land and sea within an accuracy of a few inches, thus providing information on tides as well as on storms, ice, and other phenomena. By the 1980s the system is expected to provide continuous global data.

Satellites send back an enormous amount of data by radio—so much that it can be analyzed only with the help of computers. One of the researchers engaged in this work is Edward N. Gaposchkin of the Smithsonian Astrophysical Observatory in Cambridge, Massachusetts. His mother and father are both eminent Harvard astronomers; but, instead of using a telescope, Gaposchkin employs mathematics to interpret reports from satellites.

This isn't easy. The varying gravitational pull of the earth, the sun, and the moon (depending on their relative positions) causes a satellite to move crookedly in its path; these effects have to be reckoned and accounted for in order to assess the measurements the satellite is making of tides on the earth. To do this,

might have been pulled in this direction by tidal forces, but he observed there were also geographic trends in the opposite direction and he had no confidence in his speculation. Now, eighty years later, scientists are convinced that the crust is made up of huge plates which do slide, but tidal forces are not thought to be the cause of the movement.

The sea of air, at the bottom of which we live, pulses with the movement of the moon and sun somewhat as the ocean does; but the tides are much more difficult to measure and no one was aware of them until scientists began to investigate. In the seventeenth century, René Descartes theorized that space was filled with ether and that the moon compressed it while circling the earth, thus causing tides in the sea. To test the theory, Sir Christopher Wren asked Robert Boyle, the British physicist, to make a barometer and to measure the pressure of the air as the moon passed. The barometer didn't show anything about the tides but, to their satisfaction, it did tell something about the weather.

In about 1823, Laplace tried to measure atmospheric tides with a barometer, hunting for changes in pressure — the weight of the air — that would be in rhythm with the movements of the sun and moon. He found such rhythms but decided that the range he was able to register — 0.054 millimeters — was four times as big as it ought to be; its reliability was therefore doubtful.

The first successful recording of the lunar air tide was made on St. Helena Island in 1842 by a meteorologist named Lefroy. St. Helena is in the tropics, where tidal forces are greater and barometric fluctuations lesser than in other parts of the world; elsewhere, irregular weather can cause confusing changes in atmospheric pressure. It was not until 1918 that lunar tides in the atmosphere were first demonstrated outside the tropics. The work was done at the Greenwich Observatory by Sydney Chapman, then a young researcher and later a distinguished geophysicist, who analyzed 64 years of barometric records to find the lunar rhythm, represented by a range of only 1/1000 of an inch of mercury. Through the years since then, nearly a hundred determinations of the lunar barometric tide have been made in various parts of the world, two-thirds of them by Chapman and his various collaborators.

Determination of solar tides in the atmosphere is well-nigh impossible. When the sun is overhead and is exerting its greatest gravitational attraction, it is also beaming the maximum of heat. The heated air expands and rises and as a result the barometric pressure drops, masking tidal fluctuations. A daily cycle is completed as the air cools during the night.

As we have seen, atmospheric tides at ground level are very slight. This is true of the whole troposphere—the layer of air in which we live, breathe, and fly. In the higher levels, the stratosphere, mesosphere, and ionosphere, farther from the earth's gravity and closer to the tidal forces of the sun and moon, the tides become increasingly greater. In the top layer, the ionosphere, the tide waves can be as much as three miles high.

Just as tides cause currents in the oceans, they induce currents in the atmosphere, in the form of winds. At the surface, tidal winds have a velocity of only a few inches a second but the speed increases as the density of the atmosphere becomes less. In the ionosphere, which extends from about 40 to 400 miles above the earth, the speed of the wind becomes a hundred or more times greater.

The ionosphere is made up of gases in which the sun's ultraviolet rays have stripped electrons from atoms, leaving electrically conducting ions which reflect radio and radar waves from the earth. We can study the winds by radio reflections from the drift of meteor trains, just as we can watch the contrail (condensation trail) from a high-flying jet plane drifting in the wind. A wealth of information about ionospheric tides and winds has been obtained from data obtained in studying incoherent radar backscatter with the radio telescopes at Millstone Hill in Massachusetts and Arecibo in Puerto Rico.

Because of the electrical properties of the ionosphere, magnetic tides are caused by tidal movement. Albert Defant writes:

> The system earth-atmosphere-ionosphere is looked upon as a gigantic generator with the earth as its permanent magnet. The atmosphere

acts as a coil, set in motion (a) by thermal differences caused by solar rays and heat losses through radiation, and (b) by the tide-generating forces of moon and sun. These thermal and tidal movements are like a "breathing" of the atmosphere. The electrically conducting layers of the ionosphere may be thought of as the windings of the coil. Since these move perpendicular to the earth's magnetic field, induced currents are produced, whose magnetic fields can be observed on earth.

Because lunar and solar tides cause high winds in the ionosphere, one is led to wonder if these tides may be involved in the development of weather patterns—in stirring up storms, such as those that coincide with the perigean spring ocean tides. Old weather saws say that rain is most likely at the turn of the ocean tide and that:

> If it raineth at tide's flow,
> You may safely go and mow;
> But if it raineth at the ebb,
> Then, if you like, go off to bed.

But meteorologists have found no reason to suspect that tides do influence the weather. Compared to all the other forces affecting the behavior of the atmosphere, they are minor.

Returning to the subject of earth tides, one can justifiably speculate about ponderable consequences. These will be discussed in following chapters.

NOTES

The quotation by Lord Kelvin is from his lecture on "The Tides" before the British Association on August 25, 1882. The lecture may be found in *The Harvard Classics,* Vol. 30 (P. F. Collier & Son, New York, 1910) along with extracts from a lecture to the Glasgow Science Lectures Association. I have drawn upon both for information in this chapter.

The Earth Tides by Paul Melchior (Pergamon Press, Oxford, 1966) is the definitive work on this subject. George Darwin offers some discussion in *The Tides,* previously cited. The quotation from Pliny comes from his *Natural History,* translated by H. Rackman (Harvard University Press, Cambridge,

1938), Vol. I, Book II, Chapter XCIX, pp. 343-349. I have also used as a reference source a paper entitled "An Analysis of Tidal Strain Observations from the United States of America: I. The Laterally Homogenous Tide" by Christopher Beaumont and Jon Berger in *Bulletin of the Seismological Society of America* (Vol. 65, No. 6, pp. 1613-1629, December 1975). I am grateful to Dr. Edward N. Gaposchkin of the Smithsonian Astrophysical Observatory for information and guidance.

Atmospheric Tides by Sydney Chapman and Richard S. Lindzen (D. Reidel Publishing Co., Dordrecht, Holland, 1970) is probably the best source on that subject. A Festschrift, *Sydney Chapman, Eighty* (sponsored by University of Alaska, University of Colorado, and University Corporation for Atmospheric Research, 1968) provided background.

A paper entitled "The Lunar Air Tide," by Chapman and Bernard Haurwitz (*Nature,* Vol. 213, No. 5071, pp. 9-13, January 7, 1967), was particularly useful.

The weather jingle is from *Weather Lore* by Richard Inwards (published by the Royal Meteorological Society, Rider Co., London, 1950). The authority for rain at the turn of the tide is *Eric Sloane's Weather Book* (Duell, Sloan & Pearce, New York, 1949).

IX

Grit in the Clockwork

It is neither incredible nor wonderful, if the moon, having in herself nothing corrupt or muddy, but enjoying a pure and clear light from heaven, and being full of heat, not of a burning and furious fire, but of such as is mild and harmless, has in her places admirably fair and pleasant, resplendent mountains and purple-colored cinctures or zones. PLUTARCH

IF THE MOON raises tides on the earth—even in the solid rock—doesn't the earth, which is eighty times heavier (though only four times larger), affect the moon? Indeed it does. The best evidence of the effect of earth's gravity is the occurrence of moonquakes—shudders in the structure of the moon when it is closest to the earth.

For many years scientists believed that the moon bulged toward the earth, that this bulge was a "frozen tide," and that it was the earth's pull on the bulge that kept one face of the moon toward us.

Sir George Darwin, who was nearly as important in the study of

tides and lunar evolution as his father, Charles Robert Darwin, was
to biological evolution, wrote:

> Once on a time the moon must have been mol-
> ten, and the great extinct volcanoes revealed by
> the telescope are evidences of her primitive heat.
> The molten mass must have been semi-fluid, and
> the earth must have raised enormous tides of mol-
> ten lava. Doubtless the moon once rotated rap-
> idly on her axis, and the frictional resistance to
> her tides must have impeded her rotation. . . .
> She rotated more and more slowly until the tide
> solidified, and thenceforward and to the present
> day she has shown the same face to the earth.

It was not until 1973 that the "frozen tide" was proved to be a
myth; William Kaula of the University of California, Los Angeles,
calculated the distribution of the moon's mass. Actually, he found,
the moon has a bit of a bulge on the far side. But the center of the
moon's mass is about a mile and a quarter closer to the earth than
the center of its spherical figure. According to Ursula B. Marvin of
the Smithsonian Astrophysical Observatory,

> It now appears that the offset can most easily
> be explained by the observation that the moon
> has twice as great a thickness of low-density high-
> lands crust on the far side and a greater mass of
> high-density rock in the near side.

When Orbiter 5 made eighty circuits of the moon in 1967, it
was found that the satellite slowed down over certain points, indicating that there are spots where gravitational pull is greater
than elsewhere. These spots are called "mascons" (for mass concentrations). One theory accounts for mascons by describing
them as enormous and very heavy meteorites which had plunged
into the moon and buried themselves under the surface. Maria
(lunar "seas") mark the holes made by the meteorites. Scientists
now believe that the variations in gravity are caused by the dense
basalt, which is solidified lava that poured into the maria.

The two sides of the moon are quite different. The far side is
made up chiefly of pock-marked highlands, with only a few small

The first men on the moon made this picture as they looked back from their Apollo 11 spacecraft, then 10,000 miles from the moon on its homeward journey. *(NASA)*

maria which cover only about two percent of the surface. On the face we see large maria and thousands of craters, many of them immense. The biggest of the "seas," Mare Imbrium, is large enough to contain France and Great Britain. On one side of it is a 600-mile chain of mountains, the Apennines, the highest peak of which, Mount Huygens, rises nearly 20,000 feet. The most striking of the craters, Copernicus, is fifty-six miles in diameter and has a rim as high as 17,000 feet. There have long been arguments about the presence of volcanoes. In recent years several astronomers reported having seen red flashes, but it is doubted that these were volcanic. Two possible explanations are that they were caused by puffs of emitted gas or the impact of meteorites.

Seeking to explain the difference between the two sides, John A. Wood of Smithsonian Astrophysical Observatory estimated that the leading side of the earth–moon system would have had four times as many impacts from asteroids as the "lee" side. He believes, therefore, that the heavily pitted face of the moon was once the leading side and because of its consequent density could have been pulled around to its present position by the earth's gravitation.

The heaviest bombardment of the moon is believed to have taken place between 4 and 3.9 billion years ago, though there may have been a more or less continuous rain of meteorites in earlier times. During that 100 million years, however, asteroids, as big as 600 miles in diameter, pounded into the then-hard crust. The basins which became the maria, such as Imbrium, were formed. In making craters, projectiles from space threw out chunks of debris which fell back and made smaller craters. Even on grains of dust, tiny craters are found.

The bombardment (to which the planets were also subjected) may have had other effects. William K. Hartmann of the Planetary Science Institute of Tucson, Arizona, wrote:

> The largest interplanetary bodies probably carried so much energy and momentum that, depending on the direction they approached a planet, they could have tilted it, speeded up its spin, slowed down its spin, destroyed a satellite

> or perhaps left rings of material around it after
> breaking up under gravitational forces.

George Darwin may have been wrong about the frozen tide but in seeking to explain the origin of the moon he offered an important concept — that of tidal friction. Then an astronomer and mathematician at Cambridge University, he summed up his studies in a series of Lowell Lectures in Boston in 1897, published under the title of *The Tides and Kindred Phenomena in the Solar System*. The following paragraphs recap the idea in simplified form.

Water flows so readily that it is difficult to think of friction being a factor in the tides. But Darwin felt that it was, and set out to show how to account for the friction. Darwin's approach was through what might be called the Tar Baby Effect. He studied the viscosity of pitch, which is certainly very sticky, and moved from that to consideration of what tides would be like on a viscous planet. He concluded:

> Molten rock and molten iron are rather sticky
> or viscous substances, and any movement which
> agitates them must be subject to much friction.
> Even water, which is a very good lubricant, is
> not entirely free from friction, and so our
> oceanic tides must be influenced by fluid fric-
> tion, although to a far less extent than the
> molten solid just referred to.

Darwin reasoned that tidal friction slows down the rate of the earth's rotation. A consequence of this process, thanks to the laws of physics, is that the slower the earth rotates, the farther the moon moves from the earth and the faster it revolves around the earth. If the moon has been moving farther from the earth for millions of years, there must have been a time when it was very close and it may even have been a part of the earth. Darwin calculated that a molten primitive planet (earth-plus-moon) would have rotated so rapidly that its day would have been the equivalent of only three or four of our present hours. The sun's tidal force would have been exerted in two tides a day (as is the case now) and these tides would have been only about two hours apart.

The liquid planet would have had a resonance period (like a seesaw or pendulum) of 1 hour, 34 minutes. Tidal friction would have brought the tides into closer and closer resonance with the planet's natural period and the tides would have become higher and higher (like a child pumping a swing). Enormous tides, perhaps several miles high, acted on by the centrifugal force of rapid rotation, would have resulted in the eventual detachment of material; this material would then have revolved around the earth in the form of the moon.

Other scientists elaborated on this scenario. They thought that the deep basin of the Pacific Ocean was the hole from which the moon was torn; they estimated the time since this Caesarean birth at 2 billion years. Geological evidence was offered, such as the fact that, like the earth's crust, the moon is made of relatively light materials.

This theory has been pretty well demolished. The detractors have calculated that the tidal force would have been only one-millionth of that needed to tear the moon from the earth. Studies of plate tectonics and of the ocean bottom show that the moon was not pulled from a Pacific womb. The Pacific floor is younger than 180 million years.

But Darwin was right in concluding that tidal friction slows the rotation of the earth. According to present estimates, the day becomes longer each million years by eighteen seconds and at the same time the moon moves farther from the earth. Calculating backward 1.5 billion years, this would place the moon within the Roche Limit—a distance of 11,000 miles, inside of which the earth's gravitational pull would break it to pieces. There is no evidence, on the moon or on the earth, of a cataclysmic encounter at that time, however.

Of course the rate at which the moon has moved away from the earth may not have been constant. One method of estimating the rate is to study ancient corals, which show a growth line for each time a high tide covered the little marine animals. Such studies indicate that some 400 million years ago there were 400 or more days in the year instead of the present 365, which meant that the earth was rotating faster and the moon was closer to it.

Another possible clue from the fossil record is the rather sudden appearance of hard-shelled marine animals, particularly the trilobite, at about the beginning of the Cambrian period, from 500 to 600 million years ago. Up until that time, the inhabitants of the sea had been soft-bodied, such as the jellyfish, and their fossils are rare. Then, without evidence today of immediate evolutionary ancestors, a great variety of creatures with external skeletons or shells became abundant. One hypothesis for this "eruptive evolution" is that it resulted from the arrival of the moon in orbit around the earth and the beginning of high lunar tides.

"Hard exteriors could have evolved in response to the need for protection against strong tidal currents and exposure during low tide," theorized D. L. Lamar and P. M. Merifield of Santa Monica, California. They explain:

> The feeding and breeding habits of present shallow-marine organisms are closely adjusted to diurnal and monthly periodicities in solar and lunar light intensities and tides. Before the moon became a satellite of the Earth, existing marine fauna were adjusted to a simple cycle of solar tides and light. The abrupt appearance of a lunar cycle may have led to wholesale extinctions and a period of rapid evolution to adjust to the changed conditions.

This is one of a number of arguments (not necessarily accepted) for the theory that the moon was captured by the earth. There are other, and more compelling ones, chiefly having to do with the geological record, the chemistry of rocks, the formation of lunar craters, and the behavior of orbiting bodies. Like the theory that the moon was wrenched from the earth, the capture theory has little support from scientists today. In terms of the game of Frisbee, the chief objection is that such a catch would have required a very long gravitational arm.

A third theory, which has received increasing support in recent years, is that the earth and moon were born at about the same time and in about the same area of the solar nebula—that they

130

were accretions in eddies of the great cloud of dust and gas, condensing and slowly growing as their gravity collected more material. Broadly speaking, creation was tidal. The earth may have been formed sooner, or may have grown faster, and collected more of the heavy elements. The moon must have been revolving around the earth from the very beginning or was captured shortly after.

There are difficulties in this theory, however. Two distinguished scientists, Harold C. Urey and Gordon J. F. MacDonald, reviewed the contradictory evidence and concluded facetiously that no method for the origin of the moon is possible and therefore the moon cannot exist. Yet anyone can look at the sky, and there it is, tangible, predictable, and tantalizing.

The future of the earth and moon is less controversial, perhaps because there is no conflicting evidence to worry about. Darwin theorized that tidal friction would slow the rotation of the earth until the day and the month would be the same—the equivalent of 55 present days. But then solar tidal friction would become a compelling influence. Having retreated to about 350,000 miles from the earth, the moon would now slowly be pulled back toward it. At last it would reach the Roche Limit and be broken into pieces which would revolve around the earth. Instead of a moon, we would have rings, like Saturn.

Lloyd Motz, a Columbia University astronomer, offered a dramatic picture of what would happen as the moon approached the Roche Limit:

> The ocean tides at their maximum on the earth would then be hundreds of feet high and would completely inundate all the land masses in their path as they follow the rising moon. But this would not be the worst of it, for the moon would distort the entire earth by producing huge tidal waves within the earth's rocky crust and in the underlying regions. These structural tidal waves would set off vast earthquakes and volcanic eruptions. Although the earth itself would not be destroyed by such cataclysms, all land life would be.

After the breakup of the moon, friction of solar tides would slow the rotation of the earth until the length of day would be a few weeks longer than the present year. Soon only one face of the earth would be turned toward the sun and it would become a scorching desert, with the other side of the earth dark and under thousands of feet of ice. But, Motz reassured, "These two forbidding hemispheres would be separated by a narrow zone (perhaps a few hundred miles wide) where intelligent life could exist."

Although this prognosis of what could happen 10 or perhaps 30 billion years from now seems frightening, there is really no cause to worry. Before these things could happen, a worse termination may be in store for us. The sun will heat up enough to evaporate the oceans. Then it will rapidly cool, and any water that may be left will be frozen. There will be no tides and the moon will recede to its theoretical limit. The sun will have exhausted its fuel but in a final pyrotechnic paroxysm will become a brilliant nova, melt its retinue of planets, and collapse; it will become a burnt-out star and perhaps a black hole.

Such events are in the realm of speculation — with a solid, scientific basis, of course. Hard evidence, mainly in the form of rocks brought back from the moon by astronauts, is still being studied by selenologists intent on understanding lunar (and terrestrial) history.

The oldest dated rocks among the lunar samples were pieces of dunite, an olivine from deep in the crust, which were found to be 4.5 billion years old. They were among the fragments cemented together in breccias that are between 4.25 and 3.9 billion years old; the breccias were found by the Apollo 16 crew in the Descartes Mountains, at the extreme right on the moon as seen from the earth. The mountains, or highlands, which make up two-thirds of the moon's surface and are believed to represent the original crust, may be 4.47 billion years old.

The oldest rocks found on earth have been dated at 3.8 billion years, but this does not mean that the moon is older than the earth. It only means that the earth has gone through such profound changes that older rocks have been hidden or transformed. Radioactivity calculations indicate that the earth and the moon are 4.6 billion years old and that none of the other planets are older. Birth of the universe is now placed as far back as 20 billion years and of the Milky Way, our galaxy, at 10 to 20 billion years.

If the moon was formed by the condensation of nebular material the process was not a gentle one like the condensation of dew. Gravity pulled chunks of material together with such force that tremendous heat was generated and the embryonic moon became at least partially molten. Then it cooled off, only to be heated up again by radioactivity deep inside — as in the case of the earth, between 3.8 and 3.2 billion years ago. It was at this point that lava welled up from the interior to fill the basins formed by the earlier bombardment, creating the areas thought by early astronomers to be seas — the maria. The lava solidified and probably there has been no fundamental change in the past 3 billion years.

Some meteorites continue to streak into the moon, uncushioned by an atmosphere as is the earth; they make sharply outlined craters in the maria, strewing the surface with debris from their disintegration and from the rocks they hit. The most conspicuous (to the naked eye) feature on the moon is Tycho, that bright, white spot near the bottom with rays extending almost to the middle of the disk. This crater was made only a few hundred million years ago. The rays, presumably made when a meteorite created the 54-mile-diameter pit and splashed out material, contain an abundance of aluminum; the rays are not thoroughly understood.

Copernicus, another bright crater, to the left of center, is believed by many to have been a volcano which emerged 800 million to a billion years ago. Apollo 12 landed on one of its rays and brought back a sample of the material that makes them so bright — a radioactive, ropy glass that looks like pulled taffy, coated with a crystalline dust which glitters. Because the glass is rich in potassium (K), rare earth elements (REE) and phosphorus (P), it is called KREEP. Thus far thirty or forty mineral families known on earth have been found along with three mineral species never seen before. The three were collected by the crew of Apollo 11 on Mare Tranquillitatis and were named tranquillityite, armalcolite (for Armstrong, Aldrin and Collins), and pyroxferroite.

Scientists were no more curious about the surface geology of the moon than they were about the internal structure. The moon had been generally thought to be totally cold and dead, but apparently there is a molten core about 450 miles in diameter (the earth's core is 4,000 miles across) which may be of nickel and iron or of silicate rock. Above that there is a mantle about 620 miles thick, and above that the crust is less than 38 miles thick.

Information about the interior has come primarily from four seismic stations set up on the moon by astronauts. Since the moon is a very quiet place—compared to the earth, where all sorts of things, from trucks to ocean waves, cause vibrations—the sensitive instruments can record very small events. Astronauts jettisoned their lunar modules which, crashing on the moon, caused the seismometers to send out strong and useful signals. The surface "rang like a bell" for four hours after the Apollo 13 dump.

Until the seismometers were turned off in 1977 (because of NASA's lack of funds) they dutifully telemetered back to earth the moonquake data which revealed something of the moon's structure.

A team headed by M. Nafi Toksöz of M.I.T. analyzed more than a thousand of these moonquakes and found that nearly all of them occurred at the base of the mantle where the greatest stress would be caused by the tidal influence of the earth. And they occurred in a rhythmical pattern, every twenty-seven days, when the moon was in perigee, closest to the earth and therefore subject to the greatest gravitational pull. There was another rhythm of 206 days discovered; this period corresponds to the period of a solar perturbation in which the moon is pulled slightly closer to the sun.

"The occasional near-surface moonquakes, which are not correlated with the tides, are probably due to the release at random times of the tectonic stresses frozen in the lunar crust, and they may resemble terrestrial intraplate earthquakes," the scientists observed. As to the deep quakes, "If moonquakes are associated with weak zones or faults in the lunar interior, then the tidal stress component acting on this plane would be the controlling factor in the occurrence of moonquakes."

134

Temblors on the moon are feeble but they are evidence of the power of the tides.

NOTES

The Plutarch quotation is from "On the Face in the Round of the Moon," found in de Santillana's *The Origins of Scientific Thought* and cited in Chapter II.

George Darwin's ideas on tidal friction and the origin of the moon were set forth in *The Tides*, previously cited, and also in an article, "The Evolution of Satellites," in *The Atlantic* (Vol. LXXXI, No. CCCCLXXXVI, pp. 444-455, April 1898).

I have quoted from *The Universe* by Lloyd Motz (Charles Scribner's Sons, New York, 1975), who also wrote an article on "Destruction of the Earth-Moon System" for *Natural History* (April 1976). Among the many books that have dealt with the subject, *Biography of the Earth* by George Gamow (Viking Press, New York, 1941, New American Library, 1948) is particularly readable though old.

Of the numerous books on the moon, I happened to find the following especially enlightening: *New Guide to the Moon* by Patrick Moore (W.W. Norton, New York, 1976); *The Moon Book* by Bevan M. French (Penguin Books, New York, 1977); *Pictorial Guide to the Moon* by Dinsmore Alter, revised by Joseph Jackson (Thomas Y. Crowell, New York, 1973); *Moon Rocks* by Henry S.F. Cooper, Jr., a report on the Apollo 11 conference in Houston (Dial Press, New York, 1970); *Space Science and Astronomy* by Thornton Page and Lou Williams, a report on the Third Lunar Science Conference in 1972 (Macmillan, New York, 1976); *The Moon* by Zdenek Kopal (Chapman and Hall, London, 1960). *Scientific American* articles of special interest were "The Most Primitive Objects in the Solar System" by Lawrence Grossman (February 1975, Vol. 232, No. 2, pp. 30-38) and "Cratering in the Solar System" by William K. Hartmann (January 1977, Vol. 236, No. 1, pp. 84-99). The best short summary of lunar science that I found was "The Moon After Apollo" by Ursula B. Marvin (*Technology Review*, July-August 1973, Vol. 75, No. 8, pp. 13-23). Dr. Marvin kindly and critically read this chapter.

I have quoted from "Cambrian Fossils and Origin of the Earth-Moon System" by D. L. Lamar and P. M. Merifield (*Geological Society of America Bulletin*, Vol. 78, pp. 1359-1368, November 1967) and "Moonquakes: Mechanisms and Relation to Tidal Stresses" by M. Nafi Toksöz, Neal R. Goins and Chuen Hon Cheng (*Science*, Vol. 196, pp. 979-981, May 27, 1977). Professor Toksöz and Mr. Cheng of M.I.T. also discussed the subject with me.

X

Tides and
Earthquakes

When we remember that only a thin rocky
crust, comparable to the skin of an apple,
separates us from the red-hot semi-molten interior of our planet, we do not wonder that the
inhabitants of its surface are so often reminded
of the "physical hell" lying below the peaceful
woodlands and blue seas.   GEORGE GAMOW

IF THE EARTH'S tidal pull
can cause moonquakes, can the moon's tidal pull cause earthquakes? And how about "sunquakes"?

Common sense would probably lead us to conclude that a force
that can squeeze and deform the whole earth would surely shatter
the rocky crust and cause a cataclysm. And it might make us
think that if the sun can cause tides, perhaps the moon and all the
planets can create a gravitational response in the fiery sun.

The nature of earthquakes is far too complex, however, for
common sense to be of much help. If we insist on using the word
"cause" in our questions, scientists would almost unanimously

answer "No!" But if we substitute the word "trigger" there are affirmative replies. A very small force, at the right time and place, can set very large forces in motion. A field mouse can trigger an avalanche if a large mass of snow on a mountain is poised for a catastrophic slide.

Before pursuing the subject further it may be well to review briefly the state of knowledge about earthquakes. Back when even respectable geologists thought the earth was a sort of dried apple, with mountains forming in the crust like the wrinkles in the shriveling apple skin, earthquakes were regarded as violent adjustments in the shrinking layers of rocks.

In the past dozen years the earth sciences have been revolutionized by a growing understanding of *plate tectonics* — the process by which terrestrial structure is changing. The earth's shell consists of some twelve plates which are moving like huge cakes of ice when a frozen bay or lake is broken up by the tides or winds. Some plates push under others, creating mountainous humps. In the earth, descending edges of plates melt in the hot mantle and, meanwhile, in mid-ocean floors, magma is oozing upward to create new crust. The continents, massive though they are, float with the plates and may drift apart as much as four inches a year.

Nearly all the earthquakes, and most volcanoes, occur in places where plates crunch together. One of the most active of the plates underlies the Pacific Ocean and there is violent vulcanian activity at its border — in California, Alaska, the Aleutian Islands, Japan, and the East Indies.

The American Plate, on which the North American continent rides, is moving in a direction just north of westerly and it grinds, slowly but inexorably, against the Pacific Plate at the San Andreas fault, which runs 600 miles along the coast of California. Near the fault's north end is San Francisco, which was practically destroyed by the earthquake of 1906 (San Franciscans prefer to call it the Great Fire). Los Angeles is on the Pacific Plate, near the southern end of the fault system.

It has been estimated that as many as 50,000 earthquakes occur in the world every year. Most of these, of course, are so small that they can be detected only by sensitive instruments, but there are likely to be a score violent enough to destroy a city hapless enough to be near the center. During the winter of 1811–12, an earthquake series in the Mississippi Valley, centered at New Madrid, Missouri, caused shocks to be felt as far away as Boston and Norfolk, Virginia. The Midwest was sparsely populated and the death toll, though not known, was presumably small.

On the other hand, 830,000 people are believed to have been killed in an earthquake in northern China in 1556, the greatest loss of lives in history. Nearly as many may have died in the earthquake in the industrial area of Tangshan City, near Peking, in 1976; but the government has been extremely secretive and no official figure has been disclosed.

Earthquake scientists have been making notable progress, especially in China, Japan, Russia, and the United States, on methods of predicting earthquakes, but apprehension as to the reliability of methods is heightened by nervousness about social consequences. Long-term predictions make possible such steps as strengthening vulnerable buildings, but they may also depress real estate values, hold up construction projects, cause unemployment, cut off earthquake insurance, and so on. Short-term predictions may cause panic.

In Los Angeles, 14,000 buildings have been identified as hazardous. Efforts are being made to strengthen or eliminate them before an earthquake strikes, but the task is enormous.

Discussing the improvement in earthquake prediction methods, Don Anderson, director of the Seismological Laboratory at California Institute of Technology, said,

> As far as southern California is concerned, a prediction won't really be very useful if it says in the next year we're going to have an earthquake. That doesn't tell us to do anything more than we should have been doing all along anyway—just by the simple fact that we live in an area of the world that is prone to earthquakes.
>
> However, when we get the prediction down to the point where we can say there will be an earthquake within, say, the next two days, then you can take such emergency preparations as

the evacuation of people, or at least having people leave their dwellings and sleep outside. This
is what the Chinese did. If the prediction is
vague to the extent of even a month, you're not
going to be able to save many lives. In the case
of China, perhaps, they could have ordered
people out of their houses for a whole month,
but that would clearly not be possible in Los
Angeles.

Parts of southern New England have been classified as zones of
high earthquake hazard too, but the prevailing attitude of people
who live there is it-can't-happen-here. Boston had a severe earthquake in 1755 (just seventeen days after the disastrous Lisbon
earthquake). There was relatively high seismic activity from 1725
through 1849 and then a quiet period of 125 years. After a study
of the area's history, Anthony F. Shakal and M. Nafi Toksöz of
M.I.T. noted "a marked increase in seismic energy release since
1940, which could represent the beginning of another episode of
increased seismic activity in southern New England."

L. Don Leet of Harvard wrote:

Investigators have found evidence of progressive
tilting of the ground resulting in subsidence of
the land relative to sea level at the rate of a foot
a century at Boston and about two-thirds of that
amount at New York. The restlessness of the
crust necessary for the production of earthquakes appears to be present here. . . . Lisbon
and Boston, which had their most recent earthquakes in the same month nearly two centuries
ago, are good candidates for a return to the active class within this century.

The causes of earthquakes—that is, the basic geological insta-
bilities—vary from place to place. The rocky foundations of New
England are quite different from those of California. It would
help in making predictions if one could recognize triggering
mechanisms in relation to the various structural conditions.

For a long time scientists have been looking for a periodicity
which would provide clues to when earthquakes would occur. A
Frenchman, Alexis Perrey, who kept records during the mid-

nineteenth century, claimed that quakes were most numerous when the moon is at perigee, when it is new or full, and when it crosses the meridian in any given place. Don and Florence Leet reached the same conclusion about a quarter-century ago that microtremors (they didn't say full-scale earthquakes) were a reaction of the earth's crust to such forces as the pull of the moon and other bodies, and, perhaps, to changes in the atmosphere. A Russian, G. P. Tamrazyan, studied 1205 earthquakes in the Transcaucasian area between 1917 and 1950 and reported that "there is a definite relation not only between earthquakes and tide-formation forces but also with the depth of the quake foci," the relationship being distinct for earthquakes at depths down to 250 miles at times of new and full moon.

Toksöz, who had been skeptical about such tidal effects, nevertheless came to the conclusion in 1978 that they can occur. With C. H. Cheng and Scott Hannahs, he studied 2670 seismic events in Turkey recorded between 1913 and 1970. Of these, more than 400 occurred during the last decade of the period, on the North Anatolian fault; and of this group, twice as many occurred at the time of spring tides as at other times in the lunar cycle. Of sixty intermediate earthquakes during the decade, twice as many occurred at the peak of earth tides.

Toksöz has a strong personal interest in Turkish earthquakes because he was born in Turkey and was back on a visit in 1975 when an earthquake killed 2500 in the area of Lice. Except for modern buildings, earthquake-resistant structures are almost nonexistent in Turkey and therefore methods of predicting quakes would be of vital importance.

In 1975, Thomas H. Heaton of CalTech reported on a study which "strongly suggests" that tidal stresses trigger certain types of larger earthquakes in the upper part of the earth's crust, at depths of some twenty miles or less. At about the same time, Fred W. Klein of Columbia University completed research in which he found correlations between the solid earth tide and earthquake "swarms," clusters of small earthquakes which occur in Iceland, the central Mid-Atlantic Ridge, the Imperial Valley of California, and the northern Gulf of California. Tidal triggering also appeared to control large after-shocks during nearly two years following the 1965 Rat Islands earthquake in the Aleutians, Klein reported. "In addition, sets of larger single earthquakes on Atlantic and northeast Pacific fracture zones are significantly correlated with the calculated solid tide," he observed.

There are two ways in which tidal force could trigger earthquakes. In one the force might pull apart the surfaces of a fault where the earth's crust is under stress. In the other, the force could increase the sliding movement of rock along a fault. The ocean tide could also have an influence. The heavy load of water on the sea bottom is shifted as the tides rise and fall, and the load can affect earthquake behavior. In addition, water forced through the pores of rock by compression—as on the sea bottom—can have a lubricating effect on fault lines that are liable to shift. Fergus Wood points out that if tides do trigger earthquakes (the "if" is important) the chances of their doing so would be enhanced by unusual perigean situations such as were described in Chapter 5. California earthquakes of October and November 1976 seemed to be associated with unusually high tides.

Correlation of earthquake swarms with tidal forces may be especially significant because such clusters of small earthquakes often precede large ones and might provide a signal for the prediction of disasters. Even fifty percent accuracy in prediction would be helpful, according to Toksöz. He believes dependable predictions may be possible in well-instrumented areas in the next ten years.

Meanwhile, for what it's worth, you can consult *The Old Farmer's Almanac* for "Determination of Earthquakes." There are five days, beginning when the moon "runs high" (highest above the southern horizon) which are supposed to be the time when earthquakes are most likely in the northern hemisphere. There is a similar period for the southern hemisphere when it "rides low" and there is a two-day "quake period" when it is above the equator. The Almanac's editor has failed to respond to my inquiry about the foundation of this method of prediction. Even so, it may have as much scientific validity as the prophecy that Los Angeles will be destroyed by an earthquake in 1982.

This sensational prediction was made in a book *The Jupiter Effect,* by John R. Gribbin and Stephen H. Plagemann, published in 1974. The authors did not appear to be mere dabblers in the occult. Dr. Gribbin was a member of the staff of *Nature,* the eminently respectable British scientific journal, and Dr. Plagemann was at the Institute for Space Studies at the NASA Goddard Space Flight Center in New York. They bolstered their argument with impressive references in the traditional scientific manner.

Gribbin and Plagemann believed that the moon does have a noticeable effect on the earth's crust, but their central thesis was that the planets have more influence and trigger earthquakes on a much larger scale. They arrived at this conclusion by a tortuous, and shaky, evidential path.

To begin with, the authors offered evidence that the planets create tides on the sun. The word "tides" is not used in precisely the same sense as when referring to tides in the earth's oceans. Rather it refers to gravitational disturbances—sunquakes, one might say. And the most conspicuous disturbances are *sunspots*—those enormous centers of belching energy that appear from time to time.

One source for this evidence was a study made by E. K. Bigg of Sydney, Australia, in which he undertook to correlate the sunspots between 1850 and 1960 with the orbit of Mercury, one of the smallest planets but the one closest to the sun. He reported that he did find such a correlation and that the influence was heightened when the earth, Venus, or Jupiter was on the same side of the sun as Mercury.

In 1972, K. D. Wood of the University of Colorado carried the idea further. He compared sunspot cycles with the tidal influence on the sun that was exerted by the earth, Venus, and Jupiter. His study covered the years from 1604 to 1972, and he found an impressive correlation. Wood reported that the planets' maximum tidal influence on the sun came in periods of 11.08 years, compared to the 11.05-year average period of sunspot peaks. Jupiter, which is such a large planet that it raises more than twice the tide on the sun as the earth raises, though much farther away, has an

11.86-year period of tidal influence. Wood worked out a forecast for the tidal and sunspot cycles up to the year 2104 and found a continuing correlation, with a perfect match in 1982.

Gribbin and Plagemann looked at the numbers and decided they saw an even more impressive cycle — one of 179 years which periodically brought all nine planets into positions of maximum tidal influence on the sun. They wrote:

> Each year from 1977 to 1982, as the earth moves around the sun, we will find the planets beyond Mars ever more accurately aligned. In the last couple of years first Mars and then the earth will move toward their positions in the alignment, followed by Venus. Last of all, little Mercury will spin around the sun completely four times during the year when all the other planets are lining up. Over a few critical months, there will be a superopposition with Mercury on one side of the sun and every other planet on the other, and a superconjunction with all nine planets in line on the same side of the Sun.
>
> We have already seen how dramatic the effect on sunspots can be when similar alignments involving only Mercury, Venus, earth and Jupiter occur. Is there any reason to believe that the superalignments will not produce even more dramatic effects? Certainly, dramatic effects have been expected from such auspicious events as long as man has studied the stars. Some astrologers mark the beginning of a new age by the occasion of the grand alignment — when Jupiter aligns with Mars and the moon is in the Seventh House, the Age of Aquarius begins. The Age of Aquarius will be, we are told, a time of peace and love. But will it be ushered in by a major slip of the San Andreas Fault and a wave of earthquake activity around the globe, unprecedented since seismology became a true science?

Getting away from astrology, we should turn our attention to two more substantive questions: What links are there between sunspots and earthquakes? If the planets do produce tides on the sun, thereby determining the timing and intensity of sunspots, what is the significance for us on earth?

For many years some meteorologists have sought to show a relationship between sunspot cycles and weather on earth — one theory being that sunspots reduce solar radiation, resulting in lower terrestrial temperatures and even great climatic changes, such as glacial periods. Gribbin and Plagemann cited research showing that the solar flares associated with sunspots shower the earth with bursts of cosmic rays, which not only cause magnetic storms and great auroral displays but (for reasons that are not clear) are followed by the deepening of low-pressure weather troughs in the Alaskan–Aleutian area. Low-pressure troughs bring bad weather across North America, accompanied by high winds.

The next step was for Gribbin and Plagemann to show that high winds — blowing, for example, against one side of mountain ranges — change the speed of the rotation of the earth, or, to put it differently, the length of the day. This idea is not as farfetched as it may seem. It has been shown that the moon's tidal friction is causing a long-term increase in the length of the day, and there are other influences that cause variations. In any case, Gribbin and Plagemann contended, sudden changes in the earth's spin could trigger earthquakes. An abrupt change in speed would jolt the earth, putting additional stress on unstable geologic structure.

Recapitulating the plot: planetary alignments produce tides on the sun; tide-induced solar flares influence the weather on earth; weather changes the rate of the earth's rotation; a change in the rotation can jostle geological structure; the San Andreas fault is waiting for just such a jostle. *The Jupiter Effect* concluded:

> A remarkable chain of evidence, much of it known for decades but never before linked together, points to 1982 as the year in which the Los Angeles region of the San Andreas fault will be subjected to the most massive earthquake known in the populated regions of the Earth in this century. . . . Small wonder that pieces of this

144

chain have lain about unrecognized for so long.
But now they have been put together there is no
question about the implications: in 1982 "when
the Moon is in the Seventh House, and Jupiter
aligns with Mars" and with the seven other planets of the Solar System, Los Angeles will be destroyed. The astrological link with the Dawning
of the Age of Aquarius may or may not be coincidence; that is outside the scope of this book,
which contains only scientific evidence and
reasoning.

Whether the prophecy is correct remains to be seen, but the
book certainly triggered tremors in the scientific world. Don
Anderson of CalTech was one of the scientists who had been cited
as a source of information about the relation between earthquakes and changes in the earth's rotation. He indignantly
reviewed the book for *American Scientist* and heartily damned it
as "a commercial commodity of the crassest kind." He continued,
"The authors are clearly after the large cult and astrology market
and the many Californians genuinely fearful of earthquakes."

Anderson went into some detail to refute the Gribbin-Plage-
mann thesis. He said, for instance,

The greatest period of seismic activity in recent
history started with an event of magnitude 8.7
in Japan in 1897 and continued through 1914.
This activity began during an exceptionally low
sunspot period and continued through two sunspot minima and one sunspot maximum. The
sunspot maximum was one of the mildest on
record.

He cited other weaknesses in the case and noted:

The correlation between earthquakes and rotation rate is well established, but cause and effect
are not.

Furthermore, Anderson and Emile Okal published a scientific
paper challenging K. D. Wood's correlation of planetary tides on
the sun with sunspots as "an artifact of the calculation." In other
words, the mathematics produced a wrong answer. They said:

A further look at our tidal potential values shows no drastic effect expected in 1982 when planets are supposed to align on the same side of the Sun. Indeed, better alignment will be achieved in 1990. Even then, no special tidal effect occurs because alignment of the outer planets has no pronounced effects on the tides. Alignment of the tide-raising planets within 10 degrees is a common phenomenon, occurring approximately every 10.4 years and is not associated with drastic tidal effects.

The august U.S. Naval Observatory did its part to set the facts straight. Its Nautical Almanac Office issued a diagram showing that the closest alignment in 1982, on March 10, would bring the planets within an arc of ninety-eight degrees—quite a wide spread. Further, the Office stated that, of the links offered in the Gribbin–Plagemann chain, only one can be considered "an established fact": that maximal sunspot activity increased the number of charged particles reaching the earth's atmosphere. The statement went on:

If Pluto is excluded, rough alignments of the planets Jupiter through Neptune occur approximately every 179 years. When Pluto is included, rough alignments occur at approximate intervals of 500 years. Thus the configuration of 1982 is interesting because, by human standards, it does not occur frequently. On the other hand, similar configurations have occurred many times during the history of the solar system, leaving no visible trace of their occurrence. Such phenomena may thus be characterized as interesting but apparently inconsequential.

Questions have been raised concerning physical effects that such planetary configuration might cause. The most straightforward question might be phrased as follows: Could the gravitational attraction of the aligned planets cause tidal forces on the Earth that would result in major earthquakes? Earthquakes result from

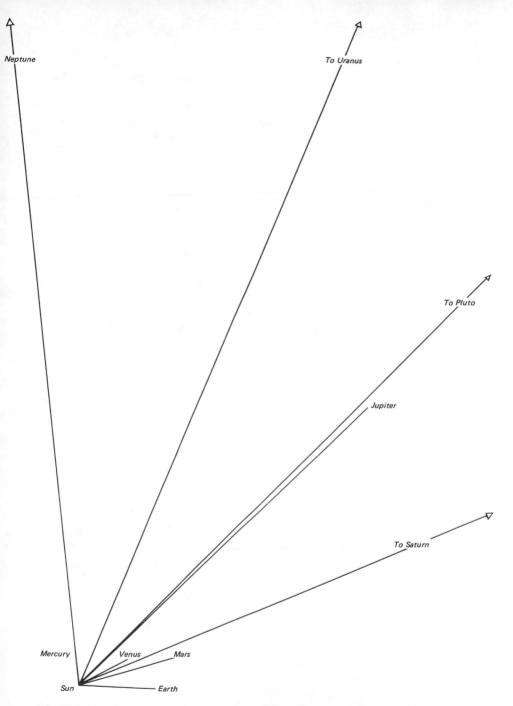

The U.S. Naval Observatory's chart showing that the "superalignment" of planets in 1982, which *The Jupiter Effect* promised would trigger great earthquakes, is not so impressive, after all. All the planets will be on the same side of the sun, but they will be spread over an arc of ninety-eight degrees. *(U.S. Naval Observatory)*

slippages along fault lines of the Earth's crust that are caused by intense pressures in the Earth's interior. It is possible, however, that internally produced stresses on the Earth's crust could reach a situation just short of earthquake conditions in which tidal forces could add just enough additional stress to bring about a quake. Thus in certain situations tidal forces might "trigger" an earthquake that was already on the verge of occurring. The principal tide raising forces on the earth are due to the Moon and the Sun, with the Moon, because of its close proximity to the Earth, having approximately twice the tide raising force of the Sun. In comparison, the tidal forces created by the planets are insignificant. If the planets were to assume the most advantageous position for raising Earth tides (a situation that will not occur in the predictable future), the resulting force would be about 0.0001 of the ordinary tidal forces due to the Moon and Sun. There is, therefore, no reason to suppose that direct tidal forces resulting from the planetary configurations of the 1980's will cause or even "trigger" earthquakes.

The major bombardment of the Gribbin–Plagemann position came from a Belgian, Jean Meeus. In an eleven-page article in the learned journal *Icarus* he cited inaccuracies, distortions, fallacies, and inconsistencies in *The Jupiter Effect*. He ridiculed the idea of a tidal effect, saying that a tidal "bulge" on the sun caused by all of the planets is only 1/2,700,000th of that caused by the moon on the earth and has been calculated at not more than one millimeter, or .039 of an inch. He concluded with these words:

> There will likely be a big earthquake in California in the near future, but it can occur as well this year as after 2000 A.D. and we have no reason to predict that the quake will take place in one specific year more likely than in another. Solar activity probably is not correlated to earthquakes, and the planetary positions certainly are not.

In reply, Gribbin and Plagemann admitted the possibility of error but did not recant. They said:

> Unlike the proponents of most useful scientific theories, however, we will be delighted if we are proved wrong and Los Angeles does not suffer a major earthquake for many years to come.

So say we all.

Before we leave the subject, however, let us return to Don Anderson's observation that there is a correlation between earthquakes and the earth's rotation rate. Nearly everyone knows that the earth does not spin on its axis with perfect smoothness. Like a spinning top, it wobbles. The axis describes a circle with respect to the stars, completing the rotation every 25,800 years; the complete rotation is known as the precession of the equinoxes. This movement is due to the pull of the moon (and to a lesser degree of the sun) on the fatty bulge around the earth's equator. Astronomers must take precession into account in determining celestial positions and astrologers regard the change from one sign of the zodiac (now into Aquarius) as significant.

There are other shifts in the angle of the earth's pole which are called *nutations*, from the Latin word for nodding. The moon precesses, and this causes a nutation cycle for the earth of 18.6 years—a period responsible for a cycle of tidal patterns which repeats itself, over and over. One nutation results from deformation of the solid earth by tides. There is a semiannual nutation as the sun moves (appears to move) north and south and a fortnightly one as the moon similarly moves back and forth across the equator. There is a seasonal nutation believed to be caused by the shifting weight resulting from variations in rain, snow, groundwater, and vegetation. (If one calculated the weight of all the leaves, grass, and pumpkins grown in a hemisphere during the summer the total would be breath-taking.)

One nutation is known as the Chandler Wobble, after Seth Carlos Chandler, a merchant and amateur astronomer in Cambridge, Massachusetts, who discovered it in 1891. Gribbin and

Papers involved in the Jupiter controversy are numerous, and I shall list only a few of them: "Sunspots and Planets" by K. D. Wood (*Nature*, Vol. 240, pp. 91-93, November 10, 1972); "Comparison of Sunspot Periods With Planetary Synodic Period Resonances" by Robert M. Wood (son of K. D. Wood), *Nature*, Vol. 255, pp. 312-313, May 22, 1975); "Earthquakes and the Rotation of the Earth" by Don L. Anderson (*Science*, Vol. 186, pp. 49-50, October 4, 1974); "On the Planetary Theory of Sunspots" by Anderson and Emile Okal (*Nature*, Vol. 253, No. 5492, pp. 511-513, February 13, 1975); Review of *The Jupiter Effect* by Don Anderson (*American Scientist*, Vol. 62, pp. 721-722, 1974); "Planetary Alignments Don't Cause Earthquakes" by David W. Hughes (*Nature*, Vol. 265, p. 13, January 6, 1977); "Comments on *The Jupiter Effect*" by Jean Meeus, "Response to Meeus" by Gribbin and Plagemann, and "Reply to Gribbin and Plagemann (*Icarus*, Vol. 26, No. 2, pp. 257-270, October 1975); "*The Jupiter Effect*" and "The Planetary Configuration of 1982," Public informational statement by The Nautical Almanac Office, U.S. Naval Observatory, February 1977; "Influence of the Planet Mercury on Sunspots" by E. K. Bigg (*Astronomical Journal*, Vol. 72, p. 463, 1967).

Scientific opinion seems almost unanimous in condemning *The Jupiter Effect* but the subject is of such widespread public interest that I feel justified in reviewing the controversy. K. D. Wood writes in a letter that he has been continuing his research on planet effects on sunspots but says (of the Gribbin–Plagemann conclusions), "I am not convinced of the correctness of their main thesis relating earthquake prediction to planet positions."

Failing to find an astrological prophecy of a new age with the planetary configuration of 1982, I wrote to John Gribbin asking for a reference. He replied, "I don't know where the 'new age' reference seeped into my consciousness." He is now at The University of Sussex, Science Policy Research Unit, Brighton, and Stephen Plagemann is at Dunsink Observatory, County Dublin, Ireland.

The Rotation of the Earth by Walter H. Munk and Gordon J. F. MacDonald (Cambridge University Press, London, 1960) is the definitive work on nutations. Ponderous though it is, the authors are not above providing a whimsical footnote on the Jules Verne and Kefauver Effects. Other pertinent references are: "Chandler Wobble, Earthquakes, Rotation and Geomagnetic Changes" by Frank Press and Peter L. Briggs (*Nature*, Vol. 256, No. 5515, pp. 270-273, July 24, 1956); "The Pole Tide" by S. P. Miller and Carl Wunsch (*Nature Physical Science*, Vol. 246, No. 155, pp. 98-102, December 17/24, 1973); "Dynamics of the Pole Tide and Damping of the Chandler Wobble" by Wunsch (*Geophysical Journal of the Royal Astronomical Society*, Vol. 29, pp. 539-550, 1974).

Tsunami

We were sleeping peacefully when we were awakened by a loud hissing sound, which sounded for all the world as if dozens of locomotives were blowing off steam directly outside the house. Puzzled, we jumped up and rushed to the front window. Where there had been a beach previously, we saw nothing but boiling water, which was sweeping over the ten-foot top of the beach ridge and coming directly at the house. FRANCIS P. SHEPARD

THE MOST TERRIFYING manifestation of the power of the ocean is what the public and press generally call a "tidal wave," though it has nothing to do with the tides. Properly it is referred to by the Japanese word *tsunami* (both singular and plural), which means "large waves in harbors." The Japanese are experts. They have experienced more than fifty in three-quarters of a century.

Tsunami have killed hundreds of thousands and resulted in millions of dollars in damage as they washed up on shores. Because they are usually caused by earthquakes and sometimes associated with volcanic eruptions, the disaster can be even greater than that produced by enormous waves alone.

Such was the case in the Krakatoa eruption of 1883 and the Thera eruption of about 1500 B.C. — probably the most devastating catastrophes in history. Some say Krakatoa was the worse of the two and, because it was abundantly documented by scientists, a persuasive argument can be made. The terror of Thera may have been greater, however, if, as others believe, the eruption was four times more powerful than that at Krakatoa.

There has been speculation that a tsunami was the basis for the Biblical story of the Flood and for similar stories from the dim past in which fact and legend are intertwined. Huge waves were not mentioned, however; Noah's deluge was produced by forty days of rain and his ark was afloat for nearly a year before it came to rest on top of Mount Ararat (16,945 feet). The Greeks had a similar story: Zeus was so disgusted with the mortals who served him a stew of human flesh and animal entrails that he decided to destroy inhuman humanity with a flood. Prometheus warned his son, Deucalion, who built an ark and rode out the storm for nine days, finally landing on top of Mount Parnassus (8060 feet). Like Noah, Deucalion sent out a dove to find land.

Looking for an event which might have stimulated such stories, some writers have selected the destruction of Thera, an island about seventy miles from Crete, which was also called Stronghyli and is now known as Santorin. The theory has also been developed, with considerable plausibility, that Thera was Atlantis, the island city described by Plato which disappeared beneath the sea.

That there was a great cataclysm is certain. Of volcanic origin, the island literally blew up, leaving a caldera of thirty-two square miles, 1,300 feet deep. Part of the remaining portion of the island was buried under a layer or pumice and volcanic ash as much as 130 feet thick. An enormous tsunami was produced by the eruption on Thera. On the island of Anaphi, fifteen miles away, it washed pumice to a height of 850 feet. It may have swept entirely over Crete, and claims are made that the wave was 300 feet high when it struck the coasts of Greece and Turkey.

This disaster is of significance because it may have brought to an end the dominance of Minoan civilization in the Mediterranean and opened the way for the flowering of Greece. Thera was

154

governed by the royal city of Knossos on Crete. According to one theory, an earthquake wrecked the palace, and the Minoan fleet, most powerful on the sea, was destroyed. The Minoans never recovered. More recently, however, some experts have come to the conclusion that the eruption on Thera and the collapse of the Minoan empire were fifty years apart, and that the former did little damage to Crete.

Whatever happened must be deduced chiefly from geological and archeological evidence. A thousand years passed before Plato wrote his account of Atlantis, and neither it nor ancient chronicles can be relied upon for detail. What happened at Krakatoa, on the other hand, was reported in many eyewitness accounts and through scientific studies.

Eruption of a volcano on Krakatoa, an island in the Sunda Strait between Java and Sumatra, started at 1:00 P.M. on August 26, 1883. There were many violent explosions, culminating in an enormous one at 10:02 A.M. the next day, when some five cubic miles of lava and ash were blown out, leaving a caldera five miles wide and 800 feet deep. The noise was heard in Madagascar, 3000 miles away.

A rain of red-hot boulders and ash caused severe damage to Krakatoa but the greatest devastation was wreaked elsewhere by tsunami from 60 to 120 feet high. The tsunami destroyed 300 towns and villages on the shores of islands, killed 36,380 people, and sank many ships. The gunboat *Berouw*, moored off Sumatra, was carried nearly two miles inland.

Tide gauges in France and Britain at the time registered a jump in the sea level and for a long while scientists wondered why this occurred. Not until 1966 was an explanation found. Frank Press of M.I.T. and David Harkrider of Brown University, with a knowledge of the effects of nuclear explosions, calculated that air pressure waves from Krakatoa, the equivalent of those caused by an atomic blast of 100 to 150 megatons, would have passed over land barriers and produced waves in the sea as far away as the English Channel.

Most tsunami are caused by seaquakes—earthquakes that fracture the ocean floor and result in the lifting or dropping of big sections of the bottom. Underwater landslides, or *slumps,* in which accumulated silt rolls down a precipice, may be involved. The displacement of enormous quantities of water sends out waves that race through the open sea at speeds of 500 or 650 miles an hour.

The waves are only a foot or two high in the open ocean and are not even noticed from ships which they pass, but as they approach land, they slow down in shallow water and build up to terrifying heights, striking a shore with a force of perhaps forty-nine tons per square yard. The first wave is often preceded by a sudden and extreme ebbing of the sea from the shore. Typically there is a series of eight waves, the third and eighth often being the largest. The usual height is from 20 to 60 feet but a scientific classification gives 242 feet as a maximum. The waves may be from 100 to 600 miles long, from crest to crest, and may arrive on shore at intervals of fifteen minutes or even an hour. Long after tsunami waves have struck there may be another series, reflected from a distant shore.

Not all earthquakes cause tsunami and the ones that do, known as tsunamigenic, are at relatively small depths in the earth — less than about sixty miles. The waves are thrown out at right angles to the earthquake fault in the sea bottom. If the waves move into a funnel-shaped bay or harbor, their energy is concentrated on a narrowing front and they grow to prodigious heights. They are not characterized by the spectacular breakers of storm waves but rather, arrive as solid walls of water.

Probably the most expert personal account of a tsunami was that given by Francis P. Shepard (quoted briefly at the beginning of this chapter). A distinguished marine geologist from the Scripps Institution of Oceanography, Shepard was staying with his wife in a cottage at Kawela Bay on northern Oahu, Hawaii, when a tsunami struck on April 1, 1946. It had been caused by an earthquake at a depth of 10,000 feet in the Aleutian Trench. A 100-foot wave demolished the Scotch Cap lighthouse on Unimak Island, seventy miles away, killing five men. Four hours later it struck Hawaii, with waves as high as fifty-six feet.

After witnessing the arrival of the first wave, Shepard got his camera, left the house, and saw the water retreating until coral reefs were exposed and stranded fish were flapping. He later wrote:

Trying to show my erudition, I said to my wife, "There will be another wave, but it won't be as exciting as the one that awakened us."

Was I mistaken? In a few minutes as I stood at the edge of the beach ridge in front of the house, I could see the water beginning to rise and swell up around the outer edges of the exposed reef; it built higher and higher and then came racing forward with amazing velocity. As it piled up in front of me, I began to wonder whether this wave was really going to be smaller than the preceding one. I called to my wife to run to the back of the house for protection, but she had already started, and I followed her just in time. As I looked back I saw the water surging over the spot where I had been standing a moment before. Suddenly we heard a terrible smashing of glass at the front of the house. The refrigerator passed us on the left side moving upright out into the cane field. On the right came a wall of water sweeping down the road that was our escape route from the area. We were startled to see that there was nothing but kindling wood left of what had been the nearby house to the east.

As the wave subsided, the Shepards hurried to higher ground, just ahead of a third and still larger wave. About six more waves came, but they were smaller. A total of 173 people lost their lives, many of them Hawaiians who went out on the exposed reefs to pick up fish and were drowned when the next wave arrived. Tsunami destroyed more than a thousand buildings and caused $25,000,000 in damage—the worst at Hilo, where a steel span was torn from a bridge and washed 300 yards upstream.

As a result of this disaster the U.S. Coast and Geodetic Survey established a Seismic Sea Wave Warning System (SSWWS), enabling seismographic and tide stations to flash warnings to the Honolulu Observatory when earthquakes and tsunami were detected. When a severe earthquake occurred along the coast of Chile in 1960 a warning was sent; again Hilo bore the brunt of a tsunami which arrived on May 22, the first wave at 9:00 P.M., a second at 12:40, and the third and largest—twenty feet high—at

1:04 A.M. A total of sixty-one people were killed and there was $20,000,000 damage. The tsunami also took 300 lives at Pitcairn Island, New Guinea, New Zealand, Okinawa, and the Philippines and 100 on the island of Honshu, Japan. Interviews with survivors at Hilo showed that nearly everyone heard the first warning but half waited until the first wave struck before fleeing from their homes and many rushed to the beach to see the excitement.

By the time of the next major tsunami, caused by the tremendous earthquake in Prince William Sound, Alaska, on March 27, 1964, the Seismic Sea Wave Warning System had fifteen seismological stations and thirty tide stations on the rim and islands of the Pacific. The network included Sitka, College, Adak, Attu, Kodiak, and Unalaska in Alaska. Again Hilo was hit but there were no casualties.

The Prince William Sound earthquake struck shortly after dusk on Good Friday, at 5:36 P.M. Alaskan time. It was one of the most violent earthquakes of modern times, with twice the energy of the San Francisco earthquake of 1906. The epicenter was only eighty miles from Anchorage, the largest city, and in an area where fifty percent of Alaska's population and all its major seaports are located. The death toll was 131, and 122 of these were victims of tsunami.

We are concerned here, of course, only with the tsunami, and they were widespread. The sea bed in one place was lifted fifty feet; this alone was enough to generate big waves. Within minutes tsunami struck many parts of Alaska and their devastation was worsened in some places by landslides which augmented the waves. At Seward, Valdez, and Whittier, the slides and waves demolished docks, railroad yards, and tank farms; oil from ruptured tanks caught fire, adding to the devastation. At Valdez, sand was splashed as high as 220 feet, and at Whittier one of the waves was 104 feet high.

Seismic waves reached the Honolulu Observatory eight minutes after the earthquake and the SSWWS went into action. The control tower at Anchorage International Airport, through which reports were supposed to be sent to Honolulu, was destroyed, and this hampered communications.

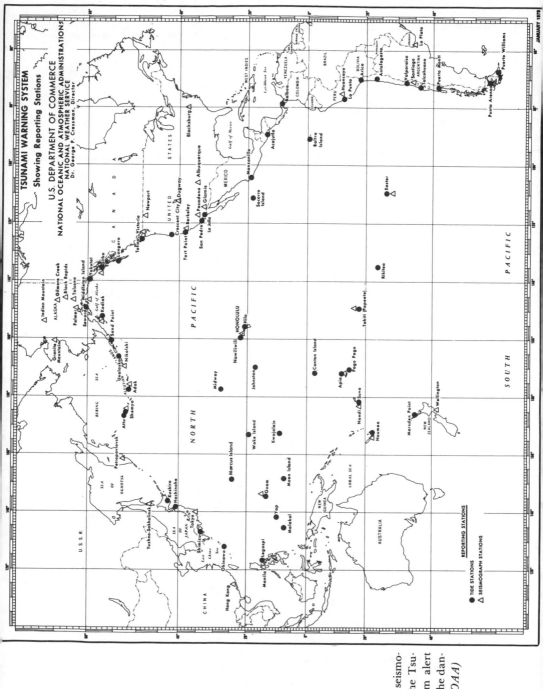

Tide stations and seismographic stations of the Tsunami Warning System alert the entire Pacific to the danger of a tsunami. (NOAA)

The earthquake occurred at 3:44 A.M., March 28, Greenwich Mean Time. At 5:02 Honolulu issued Bulletin No. 1, advising that the earthquake had occurred. At 5:30 Bulletin No. 2 was sent:

THIS IS A TIDAL WAVE (SEISMIC SEA WAVE) INFORMATION BULLETIN. DAMAGE TO COMMUNICATIONS TO ALASKA MAKES IT IMPOSSIBLE TO CONTACT TIDE OBSERVERS. IF A WAVE HAS BEEN GENERATED THE ETA'S [estimated times of arrival] ARE: ATTU 0745Z, ADAK 0700Z, DUTCH HARBOR 0630Z, KODIAK 0530Z, SAMOA 1430Z, CANTON 1530Z, JOHNSTON 1100Z, MIDWAY 0845Z, WAKE 0930Z, KWA-JALEIN 1430Z, GUAM 1315Z, TOKYO 1030Z, SITKA 0530Z, SAN PEDRO 0930Z, LA JOLLA 1000Z, BALBOA 2330Z, ACAPULCO 1300Z, CHRIST-MAS 1230Z, CRESCENT CITY 0800Z, LEGASPI 1430Z, NEAH BAY 0730Z, SAN FRANCISCO 0915Z, TAHITI 1430Z, TOFINO B.C. 0730Z, VALPA-RAISO 2200Z, HONOLULU 0900Z, HUALEIN TAIWAN 1430Z, LA PUNTA 1900Z, MARCUS 1100Z, HONG KONG 1530Z, SHIMIZU 1130Z, HACHI-NOHE 1000Z. ALL TIMES FOR 28 MARCH. [Z indicates Greenwich mean time.]

It was 6:30 when the first report of tsunami came in, this one from Kodiak in the Aleutians: Water level started rising at 0435Z. Rose 15–20 ft. above MSL. by 0445Z. Ebb tide started 0445Z. Water level 15–18 below MSL at 0507Z.

From then on the airwaves were hot with reports of the progress of the wave and reports of damage. Kodiak had only thirty minutes warning and suffered heavy losses—twenty-one lives and $31,279,000 in property damage according to final reviews. The second wave had arrived as a thirty-foot wall of water. Major damage was suffered on British Columbia, Washington, and Oregon shores.

The place hardest hit on the West Coast was Crescent City, California, just below the Oregon line, where thirty blocks were devastated and eleven people were killed. Damage was estimated at $7,414,000. Bulletin No. 2 was received at Crescent City at 7:08 A.M. (after delays in transmission through the State network) and evacuation of endangered zones was begun. The tsunami arrived at 7:39.

The first two waves caused minor flooding in the business district, and many people returned

> to the area to clean up their places of business, since past experience showed that Crescent City normally experienced one or two surges with minor flooding during a tsunami. . . .

That was the description given in an official report, which went on to say:

> This premature return to the evacuated area was the cause of most of the fatalities. The third and fourth waves caused most of the destruction and the casualties; they caught the people who had returned to the area after the second wave, and others who had failed to evacuate. Seven people, including the owner and his wife, returned to the Long Branch Tavern to remove the money from the building. Since everything appeared normal, they stopped to have a beer, and were also trapped by the third wave.

A later study by the U.S. Army Corps of Engineers stated:

> We conclude then that Crescent City's susceptibility to large wave response from major tsunami is, by its very name, related to its crescent-shaped coast and bowl-shaped continental shelf. Because of its dimensions, it will forever be a responsive echo-chamber for great tsunamis since their periods will be always capable of exciting full or partial response.

The first wave arrived at Hilo 5 hours and 24 minutes after the earthquake, which was 2 hours after warning sirens had started sounding. It was only 6 feet high and none of the seven waves that followed was higher than 12.5. Elsewhere in Hawaii the tsunami were even milder. Shores were evacuated in good time, no lives were lost, and damage was light. The rest of the Pacific also escaped disaster, though the waves went on and on, reaching Sydney, Australia in 17 hours and Bahia Esperanza in Antarctica in 22 hours and 34 minutes.

Interestingly, the earthquake caused seiches in rivers, lakes and other bodies of water in the southern states. Barges and boats broke loose from their moorings and damaged piers. The tide in the Sabine–Neches Ship Channel in Texas was 3 feet higher than normal. In a swimming pool at Corpus Christi a 2-foot wave washed 25,000 gallons of water over the edge. Artesian wells in Georgia rose 3½ feet and then fell by that much below normal. A Delaware citizen reported a 1½-inch seiche in his indoor swimming pool.

It has been estimated that there are thousands of detectable earthquakes a year, 700 of them powerful enough to cause damage, and three-quarters of them under the seas, so it is surprising that more tsunami do not occur. Although tsunami waves are not usually noticeable in the open ocean, the shock from ocean bottom earthquakes is believed to explain the mysterious jars sometimes experienced by ships as well as the uncharted reefs which sea captains have reported their ships bumped over. In 1970 the liner *Bergensfjord* was cruising in the Pacific off the coast of South America in a calm sea when there was a loud bumping sound. The ship listed and then righted itself. Later it was learned that an earthquake had occurred nearby, off the shore of Peru.

Christopher Columbus had one of his most frightening experiences on August 4, 1498, while sailing through a strait known as the Boca de la Sierpe (Serpent's Mouth) between Trinidad and Venezuela during his Third Voyage. He wrote to the King and Queen:

> Standing on the ship's deck, I heard a terrible roaring which came from the southward toward the ship. And I stood by to watch, and saw the sea lifting from west to east in the shape of a swell as high as the ship, and yet it came toward me little by little, and it was topped with a crest of white water which came roaring along with a very great noise . . . which sounded to me like the rote [roar] of surf on rocky ledges, so that today I feel that fear in my body lest the ship be swamped when she came beneath it.

The wave lifted the ship to a great height and then dropped it to near the bottom of the sea, but no damage was done. Doubtless Columbus had experienced a tsunami; though occurring chiefly in the Pacific, they are not limited to that ocean.

Some early records are vague but certainly tsunami have occurred throughout the past. In addition to those already mentioned, here is a list of major ones:

373 B.C. Helice, a town on the Gulf of Corinth in Greece was totally destroyed. An earthquake may have dropped the city into the sea.

474 B.C. Potidaea in Greece was hit by a huge wave.

A.D. 1509. The sea washed over the walls of Constantinople and Galata, Turkey.

1575. Two Spanish galleons were wrecked by a wave at Valdivia, Chile.

1692. Port Royal, Jamaica, was wrecked by an earthquake, subsidence, and sea waves.

1751. Disastrous waves hit Juan Fernandez Island after an earthquake centered at Concepcion, Chile.

1707. Tsunami killed 30,000 in Japan and swamped 1,000 boats.

1737. A wave 220 feet high hit Cape Lopatka on the Kamchatka Peninsula, Siberia.

1755. After an earthquake which destroyed Lisbon, Portugal, waves of twenty to fifty feet killed perhaps 10,000 people on the coast and washed the shores of Morocco, France, England, and the West Indies as well. (It wrecked the ship in which Voltaire's *Candide* was a passenger, providing Dr. Pangloss with another opportunity to demonstrate that this is the best of all possible worlds.)

1783. Tsunami killed 30,000 in Italy.

1793. Great loss of life occurred at Tugaru in Japan. Joseph Bernstein writes that the people "were so terrified by the extraordinary ebbing of the sea that they scurried to a higher ground. When a second quake came, they dashed back to the beach, fearing they might be buried by landslides. Just as they reached the shore, the first huge wave crashed upon them."

1820. A wave sixty to eighty feet high swept over the fort of Boele-komba at Macassar in the Celebes (Indonesia).

1868. After an earthquake in Chile, a seventy-foot wave carried the U.S. gunboat Wateree a quarter-mile inland and inundated the city of Iquique.

1896. An underwater earthquake, ninety-three miles east of Honshu, Japan, drove 100-foot waves over the beach at Sanriku when a Shinto festival was in progress. The waves killed 27,000, injured 9000 and destroyed 13,000 houses. Tide gauges at San Francisco recorded waves 10½ hours later.

1933. Again Sanriku was hit by a tsunami from an underwater earthquake: 3000 were killed, 9000 houses destroyed, and 8000 boats upset.

1958. The biggest splash in history occurred in Lituya Bay, Alaska, when an earthquake caused 90 million tons of rock and ice to fall from a glacier a thousand feet into the water, sending out a wave which scoured all trees and soil from a mountain on the opposite shore to a height of 1720 feet.

1975. Two people were killed and $2 million damage was done in Hawaii.

1976. Deaths were estimated at 8,000, with $100 million damage, on Mindanao and the Sulu Archipelago in the Philippines after an earthquake in the Moro Gulf.

Since establishment of the Seismic Sea Wave Warning System in 1948 there has been reorganization, with expansion and elaboration, under NOAA. The system is now known as the Tsunami Warning System and it involves efforts in forty-five countries on three continents bordering the Pacific, with information provided by fifty-five tide stations and twenty-four seismological stations. An International Tsunami Information Center at Honolulu is maintained by the National Weather Service. A Pacific Tsunami Warning Center at Ewa Beach, Oahu, and an Alaska Tsunami Warning Center at Palmer can provide quick action in the two most critical American areas. Offshore sensors which telemeter data to Hawaii gain some time for warnings. Satellite communications links speed the spreading of information about earthquakes and the progress of tsunami. Warnings are sent over the NOAA Weather Radio and Weather Wire systems and reach the public through civil defense and disaster offices and radio and television.

The National Weather Service emphasizes several points in its public education program. Among them:

> The Tsunami Warning System does not issue false alarms. When an ocean-wide warning is issued, a tsunami exists.
>
> Never go down to the beach to watch for a tsunami. When you can see the wave you may be too close to escape it.

NOTES

Francis P. Shepard's story of the 1946 tsunami was told in his book, *The Earth Beneath the Sea* (Johns Hopkins Press, Baltimore, 1959). I have quoted from an excerpt entitled "Catastrophic Waves From the Sea" in *The New Treasury of Science*, edited by Harlow Shapley (Harper & Row, New York, 1965).

Among other books from which I have drawn are: *The Great International Disaster Book* by James Cornell of the Smithsonian Astrophysical Observatory (Charles Scribner's Sons, New York, 1976); *Waves and Beaches* by Willard Bascom, previously cited; *Earthquakes* by D. S. Halacy, Jr. (Bobbs-Merrill, Indianapolis, 1974); *Tsunamis in the Pacific Ocean*, edited by William Mansfield Adams (East–West Center Press, Honolulu, 1970). The last is a collection of technical papers presented at the International Symposium on Tsunamis and Tsunami Research in 1969).

Exhaustive details on the Alaska earthquake and tsunami of 1964 are in the three-volume report mentioned in Chapter 5: *The Prince William Sound, Alaska, Earthquake of 1964 and Aftershocks*, Fergus J. Wood, editor-in-chief (United States Department of Commerce, Government Printing Office, Washington, 1966). Other government publications include the booklets *Tsunami! The Great Waves* and *Wave Reporting Procedures for Tide Observers in the Tsunami Warning System*.

Among magazine articles that have appeared are "Tsunami!" by Edwin P. Weigel (*NOAA*, Vol. 4, No. 1, January 1974); "Tsunami Coming" by Michael J. Mooney (*Oceans*, Vol. 8, No. 5, September–October 1975); and "Tsunamis" by Joseph Bernstein (*Scientific American*, Vol. 191, No. 2, August 1954). The last was reprinted under the title "Giant Waves" in *The World of Geology*, edited by L. Don. and Florence J. Leet (McGraw-Hill, New York, 1961). "Air-Sea Waves from the Explosion of Krakatoa" by Frank Press and David Harkrider (*Science*, Vol. 154, No. 3754, December 9, 1966) provided an interesting sidelight.

Three books on Thera were the subject of controversy a decade ago and make fascinating reading: *Voyage to Atlantis* by James W. Mavor (G.P. Putnam's Sons, New York, 1969); *Lost Atlantis, New Light on an Old Legend* by J.V. Luce (McGraw-Hill, copyright by Thames & Hudson, 1969) and *Atlantis, The Truth Behind the Legend* by A. G. Galanopoulos and Edward Bacon (Bobbs-Merrill, Indianapolis, 1969). A more recent look at the puzzle is given by D. H. Tarling in a paper, "Has Atlantis disappeared again?", a report on the Second International Thera Congress which appeared in *Nature*, Vol. 275, No. 5678, September 28, 1978.

The experience of Columbus is related in *Admiral of the Ocean Sea* by Samuel Eliot Morison (Little, Brown, Boston, 1942).

XII

Tides and Shores

It is along shorelines that the instability of the
earth's crust is perhaps most obvious, the tran-
sitory nature of scenic elements most easily seen,
and the changing relationship between land
and sea most evident.   *This Changing Earth,*
JOHN A. SHIMER

THE WINDS GOVERN
the moods and whims of the ocean — frequently tempestuous,
always restless, occasionally calm. But the tides set certain inflex-
ible limits for the ocean's behavior on shores.

As we have seen in discussing storm surges, the tides can sanc-
tion great invasions of the land. Acting with the winds, they can
destroy rocky headlands, islands, and beaches. And in changing
the contours of a shore they can also create land.

The effects of tides vary as much as the world's shores, in which
there is an infinite variety — hard granite or soft chalk cliffs,
mangrove swamps, coral reefs, sand-duned beaches, grassy salt
marshes, and so on. Where there are long fetches — thousands of
miles, perhaps — pounding waves dominate. Other factors, such
as hurricanes and monsoons, winter icing, prevailing winds, and
changes in sea level may be involved. Tides certainly dominate in

constricted waters such as estuaries, where tidal currents can scour channels or block ports with silt.

In the reshaping of shores, changes in sea level have revised the geography of whole continents. There was a time when most of the earth was covered with water. During the glacial ages, when the ice sheet reached a maximum about 18,000 years ago, much of the water was locked up in ice and sea level was reduced by 450 or maybe 520 feet. The British Isles were joined to the European continent and there was a land bridge between Siberia and North America. The New England shore extended to Georges Bank — that huge shoal where fish and perhaps oil abound — some sixty miles east of Cape Cod and extending eastward for fifty miles more. Evidence has been found there of 11,000-year-old fresh-water marsh peat, and fishermen have brought up mammoth teeth.

Melting of the ice raised sea level enough to flood great areas of the continental shelf and form bays and estuaries by drowning valleys. Authorities differ as to how rapidly sea level rose but it is believed to have come close to its present stage about 3000 years ago. As glaciers and icecaps melted it has continued to rise — possibly four inches in the past century. At the same time, the sea bottom has dropped in places because of the added weight of water. And continental areas, relieved of the burden of heavy ice, have been lifted — parts of Scandinavia as much as 1500 feet and of Canada about 900. Fossils and wave marks in highlands are souvenirs of ancient tides.

Sea level has also been changed by the slowing of the earth's rotation, which has lessened the centrifugal force so that at the equator sea level may have dropped 600 feet in the past 16,000,000 years. The massiveness of continental edges makes a difference in sea level in some places: for example, the gravitational attraction by the Andes Mountains raises the Pacific slightly near South America.

The alternation of spring and neap tides brings a short-term change in the sea level, varying the effect of surf on shores. When the tide is higher, storm waves can reach farther to attack bluffs. On a cobble or shingle beach, they not only roll pebbles up the slope but toss fist-sized rocks which build a berm, or shelf, at the surf's highest reach. One can listen to the grind and click of stones in boiling waters.

Sandy beaches are often washed down to the rocky base during the high tides and furious storms of winter. If the underwater slope is steep, the sand will be gone for good; but if it is gentle, the sand may be deposited in offshore bars. Then when summer comes, the tamer surf may wash the sand back to the shore, high tides leaving it to dry and be blown inland to build dunes.

Much sand that is washed from beaches is carried by currents along the shore—currents that are likely to be caused by waves rather than tides. Waves usually reach a shore at an angle and each one picks up sediment and moves it a bit farther along the beach.

Using fluorescent tracers, an investigator found that sand was being moved along the shore at Sandy Hook, New Jersey, at a rate of two feet or more a minute. In an experiment on the east coast of England, pebbles were dumped into the water after being given a radioactive coating so they could be traced with a Geiger counter. Even during a calm period of four weeks they moved a mean of sixty yards, and then during a stormy week some moved in the other direction as much as a mile, clear across a river mouth where there was a strong tidal current. Waves have destroyed headlands in the mid-coast of New Jersey and for the past century have been building up Sandy Hook at a rate of about 300,000 cubic yards a year. In a study of the eroding shore at Santa Barbara, California, Willard Bascom found approximately an equal rate in the transport of sand: during calm weather, a spit at the end of the Santa Barbara breakwater was growing by 250 cubic yards a day but in times of storm more than 2500 cubic yards a day were being added to it. Through the centuries, surf has been building up the outside of the hook at the end of Cape Cod and meanwhile taking large bites from the east side of the cape.

Erosion of beaches is particularly saddening to sunbathers, nature lovers, and resort operators. Transport of sand into harbor mouths may have no esthetic consequences but can be a nagging and even disastrous problem for shipping. In studying this problem, engineers use the term *"tidal prism,"* which refers to the amount of water in an estuary between mean high water and mean low water—in other words, the amount that moves in and out with the tide (including any that comes from streams).

When the tidal prism is small, and therefore ebb currents are relatively weak, waves tend to plug up a harbor entrance with sand. If it is large, the strong current is likely to build up a bar offshore, with sand arriving from littoral (seashore) drift.

San Francisco has examples of this interplay between tides and waves. The tidal prism is so large that beyond the Golden Gate there is a huge crescent-shaped bar through which a ship channel is cut. The channel requires perpetual dredging—600,000 cubic yards a year.

North of the Golden Gate, the entrance to Bolinas Bay remains relatively free because tidal and wave forces are in balance.

The ingenuity of engineers is challenged in the effort to keep sand on beaches and out of harbor entrances. Breakwaters and jetties solve the problem in some cases but in others create new problems. The mouth of the Columbia River has an extensive system of jetties but stream runoff and the supply of sand are so great, the tidal prism so large, and the waves so high, that 2,000,000 cubic yards of sand must be dredged every year to maintain a forty-eight-foot depth in the entrance channel. In some places—such as Santa Barbara, the Channel Island Harbor north of Los Angeles, and Lake Worth, Florida—sand is pumped from where it accumulates to where it is needed on beaches.

For many years a favored way to protect a beach was to build *groins*—low walls perpendicular to the coast—to slow down longshore drift and to trap sand. Groins may protect some beach-front properties but they have proved inadequate to stop drift, and they may even make matters worse. Many have been built along the New Jersey shore and estimates are that they have cut down the littoral drift by only twelve percent. In some places, the best solution has been to dump more sand on a beach. That's what is done at the world's most famous beach, Waikiki in Honolulu. Sand is trucked to the beach from dunes fourteen miles away.

While man battles with the sea over whether sandpiles of the world are to please him or please Nature, he is involved in a different kind of dispute over the control of another kind of shore — the lowland of salt marshes in the Temperate Zone and of mangrove swamps in the tropics. Usually he has been the offender.

Various processes are at work in creating marshland. The rise in sea level causes invasion of lowland. Storms can drive perigean tides higher and can flood fields once fertile; when saturated with salt these fields are able to support nothing but salt-marsh plants. Some marshes are flooded by only a few spring tides each year but that is enough to maintain their salt-marsh character.

Marshes are constantly being built in estuaries by the accumulation of silt — some conveyed by streams, some coming from erosion of shores, and some carried in by tides. Since generally flood-tide currents are faster than ebb currents, they are likely to deliver more marsh-forming material than they remove. Marsh plants take root, holding the muck and trapping sediment. The most efficient, and therefore most prolific plants on the East Coast are marsh grasses: *Spartina alterniflora*, a tall, coarse plant which thrives even where it is often flooded, and *Spartina patens*, a shorter, finer grass which grows best on higher land.

Salt-marsh grass was highly prized by our farming forefathers, when hay was the principal fuel for transportation. It would be cut and stacked on staddles — platforms of poles stuck into the mud — to cure, and would be carted out of the marsh during a low tide by horses wearing large wooden clogs to keep them from sinking into soft ground. The Massachusetts town I live in has preserved such a marsh, known as the Home Meadows, in its very center. Hay is no longer cut there but for generations the marsh has provided an open space, delighting those who watch the changing of seasons and the coming and going of ducks and herons.

A digression may be justified here for it will show in a small way what can happen to a salt marsh. The Home Meadows were supplied with salt water through the nearby Mill Pond, formed by damming the Town Brook which emptied into the harbor. The coming and going of the tide through the dam gates kept the marsh lush and turned the millstones.

171

After nearly three hundred years of service, the mill closed down. As the town grew, septic tank overflow turned the Town Brook into an open sewer and the Mill Pond became a stinking expanse of mud at low tide, decorated with old tires and tin cans. Some thirty years ago the town fathers had a bright idea. The railroad to Boston was languishing because commuters had become enamored of their automobiles. The idea was to fill the Mill Pond to provide a parking lot which could encourage commuters to use trains and thus help save the nearby railroad, at the same time getting rid of the stench.

Sewers were installed to clean up the Town Brook and when the Mill Pond was filled in, a culvert was installed to permit tidewater to reach the Home Meadows. Unfortunately, the culvert is too constricted and the meadows are not flushed adequately. Citizens complain of an odor and, lacking enough salt water, the marsh is being invaded by weeds and brush. The railroad went out of business and the town has several acres of macadam.

This example of a salt-marsh problem is trivial compared to the national destruction of tidewater assets, but it shows what can happen even when intentions are good. For decades most marshes were regarded as wastelands waiting to be exploited—at the very least they were used as dumps but additionally a large percentage were destined to be filled in for the sites of oil refineries, airports, housing developments, and the other appurtenances of civilization. The United States once had 127,000,000 acres of wetlands which have been reduced by about 200,000 acres a year until two-fifths have disappeared. Tough federal and state legislation and enforcement during the last ten years have slowed down the process. Defense of marshes and swamps by nature nuts has expanded into a public awareness and sentiment (though not always effective) for conservation.

The value of wetlands as a source of nutrition is one of the principal reasons for their protection. From sixty to eighty percent of commercial fisheries depend on the proliferation of life in estuaries. The rich muck of marshes produces vegetation that rots and is feasted on by bacteria; these bacteria serve as food for countless microscopic creatures which in turn are eaten by molluscs, crustaceans, and small fish. Marshes are the nurseries for many species. The striped mullet, for example, eats copepods, mosquito larvae, and other tiny creatures until it is about an inch long; then it browses on algae and sucks up mud to extract nourishment from detritus. Tides wash sediments back and forth, recycling the nutrients over and over.

Tidal swamps are believed to be the most productive land in the world. John and Mildred Teal calculated that "a salt marsh produces nearly ten tons of organic matter on every acre in a year" while "the best hay lands in this country produce only four tons per acre." A mussel bed on the mud flats can produce 10,000 pounds of meat per acre per year while an acre of pasture produces 100 to 200 pounds of beef per year. (Of course most Americans don't eat mussels, although Europeans adore them. Americans love clams — which Europeans don't eat — but we like beef still better.)

A South Carolina researcher found that marsh ponds can produce 100 pounds of crabs, 300 to 400 pounds of shrimp, and 250 to 400 pounds of fish per acre each year. In the Far East, where there is long experience in the intensive cultivation of marine foods, 1000 pounds of shrimp and 2000 pounds of fish per acre are produced.

It has been estimated that ninety percent of the commercial fish species in Florida depend on the food chain that starts in the mangrove swamps, which is where seventy-five percent of the nation's pink shrimp live to maturity.

The mangrove is a remarkable plant and dominates tropical lowlands as Spartina prevails in northern marshes. It thrives in tidal waters, extending a tangle of roots into the mud that hold fast even in hurricanes. The roots lift thick-leaved branches toward the sun and reach out to start new plants. The seeds sprout before they are dropped into the water and they can float for as much as a year while waiting for a shallow place to anchor. The roots exude acids which can turn a coral reef into mud to provide a home. Like salt-marsh plants, mangroves trap debris and keep sediment from washing away, so that slowly they build dry land. During the last three or four decades, mangroves are believed to have created 1500 acres at the tip of Florida.

For years, Florida developers were destroying mangrove swamps faster than the mangroves could build new land. The swamps were regarded as good for nothing much but mosquitoes. By slashing through them and dredging and filling, developers could convert a shore into a complex of high-priced house lots. All of them with a waterfront on finger-like canals.

Using this method at Marco Island on the Gulf coast, the Deltona Corporation created a suburban tropical paradise, a community with an anticipated population of 34,000. The company sold 11,000 lots. Similar developments were planned for the ten miles of shore north to Naples, and beyond. In fact, it became apparent that unless development could be restrained, the west coast of Florida would become like the east coast — its natural beauty destroyed by cooky-cutter housing, high-rise apartments, highways, and shopping centers.

Conservationists at Naples organized the Collier County Conservancy and raised enough money to create the Rookery Bay Sanctuary, with 4100 acres, which they placed under the management of the Audubon Society. Elsewhere the conservationists fought off development and worked to have federal and state regulations made stricter. Authorized by Congress, the U.S. Army Corps of Engineers promulgated a new policy which gave it the right to refuse permits for the dredging and filling not only below the mean high water line but also in areas above that line where salt water intruded. Consequently, when the Deltona Corporation sought to extend the Marco Island project into 2100 acres (in which 4000 lots had already been sold), the Corps refused permits. And a federal judge ordered the developer of a trailer park in the Florida Keys who had dredged a mangrove swamp without a permit to restore it to its natural state. At issue in such cases were thousands of acres elsewhere in Florida. Applications had been filed by various developers for changes involving 14,000 acres. The future of mangroves would depend on protracted legal battles.

Some people living in the marine Edens have found that the modest tides, admirably suited for irrigating mangrove swamps, are not very efficient in keeping sparkling, clean salt water flowing through the canal labyrinths. Polluted water can become stagnant in dead-end canals and will bloom with algae. Months may pass before rains are copious enough to flush the waterways. The only thing stirring up the water is the powerboat — in swarms — and its wake can cut into canal banks and set mud adrift.

Under these circumstances, a canal dweller may wish the mangroves were back. A Fort Myers man, M. L. Jacobs, formed a company to promote the covering of canal banks with riprap so that mangroves and other plants could take hold. He invented what he calls a "tide-pump" to help in the circulation of water through the canals.

Not the least of the values of mangrove swamps and salt marshes is their ability to help solve pollution problems. They have an amazing capacity for purifying water. John M. Teal and others of the Woods Hole Oceanographic Institution and Marine Biological Laboratory have been pouring sewage sludge on experimental marsh plots and have found that vegetation is greedy for nutrients: "Salt marsh systems are flexible enough to use as much nitrogen as is available regardless of the mechanism of supply. . . . Adequately managed salt marshes may be considered as potential tertiary treatment systems for sewage effluent." Studies in Hawaii and Fiji showed that the nitrate in sewage treatment plant effluent can be reduced by thirty percent and ammonium by sixty-three percent in passing through mangrove swamps.

After reading *Life and Death of the Salt Marsh* by John and Mildred Teal, a former University of Minnesota chemistry professor, Edgar W. Garbisch, Jr., was inspired to experiment with the restoration of marshes; he established the nonprofit Environmental Concern, Inc., at St. Michaels, Maryland, to engage in such work. He found that even on unpromising material like dredge spoil he could establish in six months the kind of marsh that it takes nature a thousand years to build. He showed how effective it is to transplant marsh plant seedlings. A homeowner on the Chesapeake Bay was losing five feet of shore a year to erosion. A stone jetty, groin, or revetment would have cost $60 per linear foot. Marsh plants were set out at a cost of $1.45 a foot and the home is now protected by a four-acre marsh. More than a hundred such projects have been undertaken in the U.S.

Lakes can be ruined by eutrophication, the process by which fertilized plants proliferate and the bacteria feeding on decaying matter deprives the water of oxygen necessary for the life of fish. But tidewaters help organisms recycle nutrients and flush out the minute plants which in still water would cause a runaway, disastrous bloom.

Flushing by tidal waters is not done with the finality of flushing a toilet, however. The process can be extremely complex and it varies greatly in different bodies of water — the chief reason being that salt water is heavier than fresh water and tends to sink beneath it when they come together in an estuary. Some inlets have rocky sills at their entrances — notably the Norwegian fjords and in some cases the inlets of British Columbia. Salt water flows over the sills and into the depths inside, where it stays, creating biologically dead waters.

A similar effect occurs at the Strait of Gibraltar, where there is a sill. The Mediterranean is very salty because of the high evaporation rate: on the ebb tide its extra-salty water flows out over the sill and sinks to depths of some 4000 feet in the lighter ocean water. On the incoming tide, regular ocean water flows over the top of Mediterranean water at the sill. Before oceanographers were able to analyze the situation, observers wondered why the tide always flowed into the Mediterranean but never seemed to flow out. Because of its salinity and warmth, Mediterranean water can now be traced for thousands of miles in the ocean.

In estuaries, tidal water sometimes mixes well with the fresh water, sometimes forms stratified layers, and sometimes forms a wedge underneath it, carrying sediment — and pollution — upstream. Tides in the Hudson River go all the way to Troy — 150 miles. Although there is a net runoff from the river, depending on the seasons and the rains as to the extent, communities that pour sewage into the stream cannot count on its being flushed right out to sea, gone for good. It does not flow all the way up to Troy with the tide, but there is a washing back and forth; pollution in New York's Upper and Lower Bay may be washed back into the river.

An estimate has been made that a particle in the water would move eight miles up the river on the flood and eight miles down on the ebb if the current were the same in both directions. Fortunately, the ebb flow is greater. But this applies only to a particle—or a discarded grapefruit rind—a single thing floating on the surface. The Hudson is what is termed a "well-mixed" stream as it flows past Manhattan, and pollution in it does not behave as simply as the grapefruit.

There is an old saying that "Running water purifies itself." There is truth to this, for filth is diluted when well mixed with water, decomposed, consumed by microscopic organisms, and carried out to sea. But decomposition requires oxygen; if sewage is mixed with water that is already polluted and deficient in oxygen, the purifying process doesn't work very well. New York City has been producing 1.3 billion gallons of sewage a day, including 360 million gallons of raw sewage; the area into which it drains, some twelve miles beyond the harbor, is considered a "dead sea." The running water of the tides and of fresh water can hardly be expected to purify all that.

Yet improvement in the situation is being accomplished. New York City is building a billion-dollar sewage purification plant, to be completed in 1986, and cities and industries upstream have reduced their contributions to pollution. For example, the International Paper Company at Corinth in upstate New York, which used to be a major polluter—pouring 24 million gallons of waste water per day into the Hudson—has installed a three-million-dollar treatment plant and has cut its use of water by two-thirds.

Organic wastes can be degraded but there are toxic substances that threaten permanent contamination. These include pesticides, such as DDT, which make their way up through the food chain, and heavy metals—notably mercury, lead, arsenic, cadmium, chromium, and nickel—which are cumulative poisons. The use of DDT has been curtailed but there are still concentrations of it in some waters. The plume of the Columbia River can

177

be traced for 250 miles along the Oregon coast by the presence of chromium in the surface water.

PCBs—polychlorinated biphenyls used for insulation in electronics equipment—are another kind of deadly poison that affected 1500 Japanese and killed quantities of fish and birds before the danger was realized. Monsanto Company, the only manufacturer of the chemicals, stopped making them and General Electric Company, the major user, stopped discharging them in waste water. There are still believed to be concentrations totaling 230 tons of PCBs in Hudson River sediments. Fishing in the Housatonic River, which empties into Long Island Sound, had to be banned because of the PCB residue from the GE plant at Pittsfield, Massachusetts. Lobster fishing was stopped along a twenty-mile stretch of the shore at New Bedford because of the chemical's presence in sediment from the Acushnet River.

Oil spills are still a menace to estuarine waters and will continue to be—no matter how stringent the government regulations—as long as tankers have to deliver their cargoes at ports and are susceptible to the hazards of the seas.

However gloomy the picture of pollution appears, the outlook is not hopeless. The Hudson River is said to be much cleaner than it was a decade ago and fish stocks there are recovering. The condition of other streams is being improved as well.

Take the case of the Merrimack River, which has been one of the most abused streams in the nation. Its sources are in the White Mountains and it empties into the Atlantic in Massachusetts at Plum Island, behind which the tides bathe a great expanse of salt marshes that once provided some of the best waterfowl hunting in the country.

The Merrimack was an important source of fish for the Indians and in colonial times it was still bountiful. A 1762 diary recorded that 2500 shad were netted in a single haul. One man who began fishing in 1789 recalled later that it was common for him to seine 60 to 100 salmon in a day.

Dams for irrigation and waterpower blocked the Merrimack to fish migrations, and the salmon disappeared by 1859. After the Civil War, attempts were made to restore the fishery. Fishways were built to enable spawning migrants to pass the dams. A million shad eggs were brought from the Connecticut River and 20,000 salmon eggs from the Miramichi River. There were other stockings through the next few decades. Runs of adult salmon resulted but pollution, overfishing, drought, and ineffectiveness of fish ladders doomed the attempts of restoration by the turn of the century.

Very little was done for the Merrimack for sixty-eight years, until after the passage in 1966 of the Anadromous Fish Act and other legislation providing funds for research and improvement. Installation of sewage treatment plants by various communities has reduced pollution. New fishways were designed and a number of them will have been built by 1985. Smolt-stocking facilities made possible the release in Massachusetts of 2500 young Atlantic salmon, ready for the sea, in 1976, about 25,000 in 1977 and 30,000 in 1978. After two years in the ocean, some from the first class were due to come back to spawn. Earlier, one thirty-one-inch egg-laden female showed up at a dam in Lawrence, Massachusetts, in November 1968, and soon expired.

Maine was doing better. Back in 1927 the Penobscot River yielded 354 salmon and then they became scarcer and scarcer. In 1970 only one was taken. But after efforts had been made to clean up the river and a new fishway had been built at the Bangor Dam, and after heavy stocking, the biggest one-day catch in modern history occurred in June of 1978—when twenty-nine salmon were taken from the Bangor–Brewer pool. By the time the best part of the season was over, the total was 285, the largest weighing eighteen pounds. But to make possible such an accomplishment, 300,000 smolt had been stocked, and the stocking rate has continued at a rate of 200,000 a year. There is a fear that nature will continue to need this aid. The Connecticut River produced eighty-six salmon during the best part of the 1978 season, an encouraging sign but not impressive enough to show that the battle for the restoration of the river had been won.

Most Atlantic salmon from American waters spend their youth feeding on shrimp and capelin off the west coast of Greenland before returning to their native streams to spawn. This is the reverse of the curious plan of the eels, in which there is an increasing commercial interest because of foreign demand.

An old New England winter sport is to gig for eels, thrusting a pronged spear through a hole in the ice to impale them as they lie hibernating in the mud. On a bitterly cold day an eel pulled out onto the ice is quick-frozen and there is no problem in wrestling with the slippery creatures as one must do in the summer.

The eels under the ice in salt water are males. The females live in the fresh water streams and ponds (they can travel overland to reach the latter) until they are sexually mature—in six or eight years. Then in the autumn, when streams are swollen and tides are high, they travel down to salt water, to be joined by males for a trip to the ocean.

Exactly what happens then is not known, for no eel has ever been caught off the American coast. But the destination for millions of eels is the Sargasso Sea, that enormous area—as big as the United States—in the North Atlantic. It is there that the larvae of the eel are found. They are leaf-shaped, transparent creatures, only about a third of an inch long, so unlike eels that when they were first found by a German naturalist in 1885 they were thought to be a new species of fish.

Further investigation has shown that there are two populations of eel larvae, one representing the American branch of the family and one the European. The former spend a year in the slowly circling waters of the Sargasso gyre, turn into elvers—eelish in form, and migrate to estuaries and streams in the spring. The European larvae take two and a half years to make a circuit of the Sargasso Sea and return to their ancestral waters. Presumably parent eels die in the Sargasso after spawning.

A supposition is that the eels began their migratory pattern some 200,000,000 years ago when the continents of Europe and America were joined, or were close together. As the continental plates moved farther and farther apart, the eels continued to travel to the central nursery and the baby eels went back to their respective homes without ever learning the difference. But the theory does not settle a momentous question: How do the eels know where to go and how to get there? Science has a lot to learn before it can answer that question.

Less mysterious is the movement of the anadromous species, the fish that spend most of their lives in salt water but return to the rivers of their birth when they are ready for parenthood. One of the most exhaustive investigations of recent times was made of the shad, largest member of the herring family, under the direction of William C. Leggett of the Connecticut River Ecological Study. More than 32,000 shad were tagged and released and better than 5000 of them were caught again to gain knowledge of their migrations. This and other studies have shown that the shad move northward along the Atlantic Coast, entering natal rivers as spring weather progresses. They spawn in the St. Johns River in Florida in January and reach the St. John River in New Brunswick in June, with contingents spawning in intermediate estuaries. They move southward again in the fall and winter as the Atlantic gets colder. They seem to seek waters where the temperature is from fifty-five to sixty-five degrees Fahrenheit.

But how do the shad recognize the rivers to which the various clans belong? Ultrasonic tracking near the mouth of the Connecticut River showed that tidal currents, salinity, and temperature provided the landmarks. In the open water, tidal currents seem to show the way. The shad tend to orient themselves toward the currents. The tidal currents in Long Island Sound at the approach of the Connecticut reverse direction every six hours. When the tide is coming in from the east, the shad swim against it but only fast enough to hold their position. When the current is from the west, on the outgoing tide, the shad swim faster and make progress toward the river — their progress becoming greater as the salinity drops and the temperature increases.

"It may be that the shad are actually responding to chemical substances characteristic of the Connecticut River water," Leggett wrote. "Indeed, this is the most likely hypothesis, inasmuch as changes in temperature and salinity alone would not be adequate clues for recognizing a specific river."

To determine which senses were more important, some of the

shad were deprived of their ability to see and some of their ability to smell; and some were deprived of both sight and smell. Loss of either sense reduced the ability to find the river but loss of both made finding it impossible. So apparently both seeing and smelling enable shad to find their way.

The Connecticut River has tides for forty-five miles from its mouth but it has a large discharge and a salt-water wedge below the fresh water extends only a few miles upstream. When entering the river, the shad move back and forth with the tides, apparently while adjusting to low salinity, but once in fresh water they proceed upstream, regardless of the direction of the current. They spawn at night near sandy bars and shoals. The eggs are fertilized at the surface of the water, drift downstream, and hatch in three to eight days. When the young are big enough to swim against currents and when the temperature drops below fifty degrees, they head for the sea, where their parents already are migrating toward the south. After four or five years in the sea the new generation will make their first trip to the rivers.

Shad are the favorite prey for some fishermen but most anglers of the Atlantic estuaries regard them as plebeians compared to the striped bass, the record for which is 72 pounds and 4 feet, 6½ inches in length. In size and fighting characteristics, the striper is less noble than the offshore game fishes such as the marlin, tuna, and sailfish. It is an aristocrat, however: abundant enough at times, though often scarce and elusive, and varied enough in feeding habits to challenge piscatorial skills. In fact, it may take on the aspects of Moby Dick and lead a man into a lifetime obsession.

The striped bass breeds chiefly in large estuaries, notably the Chesapeake Bay system, spawning above salt water in late May or June. Newly hatched fry migrate vertically. At night they come to the surface and are carried downstream in fresh water. During the day they move to the bottom and are carried upstream by the tidal current in the salt-water wedge, which is rich in plankton.

Adult fish migrate, with the arrival of warm weather, from the southern Atlantic as far as the Gulf of Maine, returning south in the fall. Cuttyhunk, the most famous center for striper fishing, gets them going and coming, with the last of the fishing in mid-

November. But apparently there are small local populations that remain in the north year-round. A horror story is told of how commercial fishermen used to catch them during the winter in the Parker River, north of Boston. They would cut a channel in the ice across the river and place a net in it. Then, upstream above what was known as a congregating point for bass, they would drop a can of acid through a hole in the ice. Blinded and gasping, the stripers would swim into the waiting net.

Commercial fishermen still catch large numbers of striped bass by seining in waters near the shore, to the abiding fury of sportsmen who sometimes go through an entire season without catching a fish. Declining catches in recent years indicate that severe damage has been done to the species.

Of course, like other fish, stripers have cycles of abundance, probably explained in part by the variations in the complex food chain of which they are a part. They feed on squid, menhaden, mackerel, shrimp, eels, seaworms, and various other prey—which means that the fisherman must use great variety in baits and methods. Surfcasting, from the beaches of New Jersey, Long Island, and Cape Cod is the most dramatic form of the chase and has produced some of the biggest fish; but trolling is a more reliable method, for striped bass love to feed in the turbulence of tidal rips and usually the only way to reach such places is by boat. When the stripers are chasing bait fish to the surface and gulls and terns are diving for them from overhead the scene is enough to drive a fisherman into hysterics.

The fanatical surfcaster may fish all night, perhaps preferring the time of a full moon when the plugs that he casts will be more visible and the tide higher. Two hours before and two hours after high tide, and never during the slack, are generally regarded as the best time for fishing, though there is disagreement.

Spring tides bring "gamefish inshore to feed, particularly in rich, marshy backwaters they can't reach during ordinary high tides," writes Joseph D. Bates, Jr., in *Fishing*.

> Fish on spawning migrations seem to expect these tides, waiting in estuaries for them so they can pass over tidewater obstructions otherwise

insurmountable. Fish see lures and baits easier during full moon and baitfish seem to travel more actively then. All this means that extra-good fishing can be expected during the high halves of tides when the moon is full and when the moon shows a thin crescent. Conversely, the neap tides during the moon's quarters may not be as good as average, but there may be holes and other hot spots such as outer bars which only can be reached then. Very low tides concentrate the fish in channels.

A different view is this: "Morning, noon or night — high tides, half tide or dead low — stripers flurry when conditions please them." So say Henry Lyman and Frank Woolner in *The Complete Book of Striped Bass Fishing*. They go on:

> Just as the angling fraternity believes it has discovered the exact time (double-checked by moon phase, wind direction, barometer and tide) to fish, these striped furies change their feeding schedule. We remember one of those unexpected rampages: it erupted at high noon of a hot August day — and the tide was flat ebb!"

And I remember a frenzied episode when pogies (menhaden) were leaping out of the water to escape the slashing jaws of scores of bluefish, and the tide was so low that in my excitement I grounded the boat.

The one sure rule is that hungry fish go where they can find food.

NOTES

John A. Shimer's quotation is from *This Changing Earth* (Harper & Row, New York, 1968). Other books from which I drew information included *Waves and Beaches* by Willard Bascom, previously cited; *Coasts* by E. C. F. Bird (M.I.T. Press, Cambridge, 1969); *Physical Oceanography of Estuaries* by Charles B. Officer (John Wiley & Sons, New York, 1976); *Western North Atlantic Ocean* by K. O. Emery and Elazar Uchipi (American Association of Petroleum Geologists, Tulsa, 1972); *The New World of the Oceans* by Daniel Behrman (Little, Brown, Boston, 1969); "The Role of Man in Estuarine Processes" by L. Eugene Cronin in *Man's Impact on Environment*, edited by Thomas R. Detwyler (McGraw-Hill, New York, 1971); "Tidal Dynamics in Estuaries" by Donald R. F. Harleman in *Estuary and Coastline Hydrodynamics*, already mentioned.

A special issue of *Oceanus* on estuaries (Vol. 19, No. 5, fall 1976) included eight articles of interest. Also on estuaries, "The Coastal Challenge" by Douglas L. Inman and Birchard M. Brush (*Science*, Vol. 181, pp. 20-31, July 6, 1973) gives a comprehensive view.

Life and Death of the Salt Marsh by John and Mildred Teal (Atlantic–Little, Brown, Boston, 1971) is both informative and fascinating, as is their *The Sargasso Sea* (Little, Brown, 1975), from which the story of eels was chiefly drawn. *How Wildlife Survives Natural Disasters* by Sarah R. Rieman (David McKay Co., New York, 1977) and *The Living Dock at Panacea* by Jack Rudloe (Alfred A. Knopf, New York, 1977) are out of the ordinary.

"Mangrove Island Is Reprieved by Army Engineers" by Don Moser (*Smithsonian*, Vol. 7, No. 10, January 1977) and "The Tree Nobody Liked" by Rick Gore (*National Geographic*, Vol. 151, No. 5, May 1977) are excellent reviews of the mangrove battle in Florida. Harrison D. Ford, Jacksonville District Counsel of the Army Corps of Engineers, supplied additional information in 1978. *Environmental Canal Systems Bulletin* (Fort Myers, Fla., October 1975) contained the account of the tide-pump. *No Farther Retreat: The Fight to Save Florida* by Raymond Dassman (MacMillan, New York, 1971) was an indispensable source.

"The Hudson: 'That River's Alive' " by Alice J. Hall (*National Geographic*, Vol. 153, No. January 1978) is a heartening article, as is "The Merrimack River" by Peter H. Oatis (*Massachusetts Wildlife*, Vol. XXVII, No. 3, May–June 1977). Information on shad research came chiefly from "The Migrations of the Shad" by William C. Leggett (*Scientific American*, Vol. 228, No. 3, pp. 92-98).

The two views on where and when to catch fish came from *Fishing* by Joseph D. Bates, Jr. (Outdoor Life, Popular Science Publishing Co., New York, 1973) and *The Complete Book of Striped Bass Fishing* by Henry Lyman and Frank Woolner (A. S. Barnes, New York, 1954). *Spindrift* by John J. Rowlands (W. W. Norton, New York, 1960) is recommended to lovers of the seashore.

Information about restoration of salt marshes can be found in "Revegetation and Development of Tidal Marshlands" by Edgar W. Garbisch, Jr. (Proceedings of the Soil Conservation Society of America, August 7-10, 1977). The same subject is treated in "Recent and Planned Marsh Establishment Work Throughout the Contiguous United States," a report by Garbisch about Environmental Concern, Inc.; this report was prepared for the U. S. Army Corps of Engineers. Another good article is "The Man Who Makes Marshes" by Robert H. Boyle, (*Sports Illustrated*, October 20, 1975).

XIII

Tides of Life

The full moon half-way up the sky
And Orion hunching in the East
With promise of months of cold and snow,
On a night so still and clear, for all
The brightness of the moon the stars
Swayed in the water, with the islands,
Back-lit, were deep as mouths of caves,
There came, through the closed windows
And doors, over the minor talk,
A torrent of gulls' cries, wave on wave,
Mounting in outrage, horror and despair,
Thousands together.

Their cries like ambulance sirens pulled me
Outside and down the hill to the sea-wall,
To strain into the darkness until
At last I understood their message:
They'd just discovered — who knows how —
The tide was never going out again,
Their favorite flats would be forever
Submerged, and they would slowly starve.

Turns out that they were wrong about
The tide, but just because it had
Gone out before was flimsy ground
For hope, especially on such
A night, surrounded by such cries.
                    Gulls BY E. A. MUIR*

AS THE EASTERN sky lightens at the approach of dawn, birds begin to stir. A rooster crows. The cows, ruminating in a corner of the pasture, start grazing on the dewy grass which, as the sun rises, resumes its task of converting solar energy into the tissues of growth. The farmer gets up and goes about his chores. In the city, people who see only patches of sky return to the routines of the day. Most lives are on a twenty-four hour schedule — which scientists call a circadian cycle — even those of nocturnal animals and nocturnal humans. All nature seems to be in tune with the sun.

But not all nature. The moon is the ruler of a remarkably large number of living things — chiefly those that dwell on the margins of seas affected by tides. As the waters retreat across beaches and mud flats, a wondrous array of delicacies is offered to shore birds. Sandpipers frantically dart along just beyond the wash of waves, gorging themselves on sand fleas. Herring gulls, which have been in dutiful attendance at the town dump, return to their natural habitat and search for stranded starfish. (Geometry says it is impossible for a gull to swallow a five-inch starfish, but gulp them down they do.)

Blue mussels, which anchor themselves with tough cords in thick mats on the tidal flats, close their shells tightly to preserve their moisture while exposed to the sun and doubtless would think, if they could think, that they are safe from predators. But a hungry gull can tear a mussel loose from the mat and, soaring a few yards, drop it on the rocks to crack the shell and expose the soft contents.

Possibly no bird is more versatile in food gathering and less choosy in what it eats than the gull. It can sail for hours in the wind wake of a ship, waiting to pick up garbage. It can hover over the water and dive unerringly for small fish. The egrets and other herons and cranes follow a different strategy. They stand, frozen,

Gulls, by E. A. Muir, copyright © 1965 by *Harper's Magazine.* Used by permission.

in tidal shallows until a silverside ventures close, and then stab with their long bills to catch it. When the tide is low, the merganser is within easy diving range of bottom-dwelling fish. (A merganser with a young flounder in its mouth has the same problem as a gull with a starfish, but it manages.)

Traditional theology might reason that the Lord created low tide to put oysters and clams at the disposal of man. Whatever the Lord's intentions, sea creatures who are dwellers of the intertidal zone have spent millions of years adapting to harsh changes in the environment—they must endure the alternation of benign seawater with unbreathable air, which may be scorching hot in the summer and icy in the winter.

Adaptation to such changes has removed intertidal organisms somewhat from the fierce competition of the sea. But it has made them vulnerable to such predators as gulls and clamdiggers. As far as we know, however, at this stage of evolution the price is not too high. They occupy a niche in which they can thrive—unless there is a radical change in the environment. That is a possibility, thanks to human pollution, but at least there will be no marked change in the tides.

The ability to exist in air as well as sea water has transformed the character of some creatures such as land crabs and certain molluscs. The European rock periwinkle is almost independent of the sea, releasing its eggs when spring tides reach its habitat.

That man's ultimate place of origin was the ocean is not doubted. His federation of cells still maintains an environment of salt water, though he has advanced considerably beyond the jellyfish. (A cynic could find similarities, in the human tendency to drift with the current and to sting prey.) Whether the acquisition of lungs and legs by man's ancestors at a much later date involved emergence from sea water or fresh water is a subject of some controversy. The genealogy is very confusing. Experts on evolution tend to favor tideless, shallow inland seas as our ancestral home. In a period of great aridity, some 400 million years ago, such seas were shrinking. The lung fishes survived in stagnant, oxygen-depleted pools by coming to the surface and gulping air. In another 100 million years our forebears had acquired legs. Mobility plus air breathing were important advantages but these animals still returned to the water to lay eggs, and the young were born with gills. Such were the *Steocephelia*, armored creatures looking somewhat like bob-tailed alligators, which were abundant in the swampy forests of the Mississippian period when beds of coal were laid down.

Although intertidal zones — rocky shores, sandy beaches, muddy flats, and marshes — may not have been important to our own evolution, they have been cradles for other organisms tolerant of twice-daily changes in the environment. The strategies for survival have been varied. One could hide in the moist muck, as do many kinds of worms and countless microscopic species. Or grow shells, like oysters, clams, and snails. But above all, success required an enormous capacity for procreation, the ability to produce eggs by the millions and to fertilize them. Fecundity provided good odds for community survival (though poor ones for the individual) and also contributed large quantities of larvae to the plankton supply, which most marine life feeds on, directly or indirectly.

Once made, an adaptation was perhaps too successful for further evolutionary experimentation. Consider the barnacle, which, throughout the world, studs practically every hard surface washed by the sea — rocks pounded by surf in which nothing else could live, pilings in polluted waters, the bottoms of boats in spite of poisoning paint, and the backs of whales and horseshoe crabs.

Barnacles belong to the *Crustacea,* which include lobsters, crabs, and copepods, the tiny members of the plankton population which swarm through the sea like insects. Infant barnacles are part of the throng, but when one finds a vacant spot on a rock, it cements itself there and builds a shell of lime, with a trapdoor which it can close tightly at low tide. During flood tide the door is opened and, with its feathery limbs, the barnacle sweeps up copepods and other foods. Thomas Huxley described it as an animal that lives by "kicking food into its mouth with its legs."

The acorn barnacle is the commonest kind. The name "barnacle" originally was applied only to the goose barnacle, which grows on a long stalk or neck and got its name from the bernacle, a northern goose which winters in England. According to medieval myth, the goose existed in infancy as this marine animal; the long neck and feathery legs were evidence enough. There was theological dispute over whether a goose was actually a fish and therefore could be eaten on Friday.

But barnacles have more important business than giving birth to geese. They are hermaphrodites, although tiny males live parasitically in some shells; and they produce large numbers of eggs, to be carried by the tides to new homes. The chance of a larva drifting to a suitable homestead is very small, and therefore the progeny must be abundant.

The oyster spat, which to survive must attach itself to a hard object where it won't be covered with silt and where strong tidal currents will bring it plenty of food, has the same sort of problem. The mother's solution is to produce 60 million eggs each season, but before they hatch they must be lucky enough to encounter drifting sperm. In an effort to prevail in this lottery, oysters begin life as males, later are hermaphrodites, then become females, and eventually may be males again.

Before facing the ultimate hazard — being gulped down by man — oysters must survive such dangers as disease, oyster drills which bore holes through the shells and suck out the contents, starfish which force open the shells, and crabs and drumfish which crack them to get inside. Barnacles, clams, and mussels have the same or similar enemies. In fact, every animal and plant living in the sea fits into the food chain somewhere.

At the bottom of the chain are the plants, chiefly algae, which collect energy from the sun and pass it on to the organisms that feed on them. One single-celled alga, *Hantzchia virgata,* which is found on the north shore of Cape Cod, has made a perfect adaptation to the tides. It retreats beneath the sand during high water, thus hiding from most browsing marine animals. But when the sea ebbs, it comes to the surface in such multitudes that it forms a "golden brown carpet," in the words of John D. Palmer of the University of Massachusetts. That may not seem so remarkable, because most plants seek the sun, but when a colony was transferred to the Marine Biological Laboratory at Woods Hole and kept under constant light and temperature, the algae continued to hide and emerge on the schedule of the tides. This diatom has no brain and no eyes, so how can it keep in tune with the moon?

Just as mysterious are the ways of many other creatures. Fiddler crabs, which abound in the muddy channels of salt marshes, retreat into burrows when the tide is in, hiding from marine predators. When the tide is out, they emerge; then the male, during breeding season, beckons to passing females with his fiddle-sized claw, urging them to visit his burrow. Taken to the laboratory, fiddlers stay on the same tidal timetable for as long as five weeks. Returned to the marsh later, they get back on the lunar schedule. Another crustacean, called the penultimate-hour crab because it becomes most active just an hour before midnight, is apparently synchronized with the solar day but its activity also varies with the moon, being greater during flood tides. Green crabs are most active at high tide.

The sea world has many other organisms that go by the tide clock in feeding or spawning. On the East Coast, *Melampus*, an air-breathing snail, climbs the stalks of marsh grasses as the tide rises. It breeds a day or so after a spring tide and then again after the next spring tide two weeks later. In the laboratory its behavior still complies with tide tables. *Excirolana*, the sand hopper or sand flea, in Southern California, burrows in the sand until high tide.

The shanny, an English member of the blenny tribe of small fish, which has the capacity of living above the waterline, was studied in an aquarium and was found to creep out of the water and lie on a rock at the approach of time for low tide.

In Bermuda, the chiton, a mollusc which looks like the sowbug, spawns only after sunrise when the tide is rising after an early morning low tide. A sea urchin in the Tortugas spawns only on the night of the full moon. There is a hydroid on the British coast that releases the medusae—the jellyfish offspring in its life cycle—only during the moon's third quarter.

Most famous of the moon-directed creatures are the palola worms, dwelling in reefs of the South Pacific and West Indies. Periodically the worms swarm to the surface, and since they are luminous, they create quite a display at night. One theory about the mysterious light that Columbus saw on the night of October 11, 1492, four hours before land was sighted, is that it came from the palola worms. (A natives' bonfire is another explanation.)

The palola worm reproduces by shedding its egg-bearing tail, and while the front end returns to a hiding place in the rocks, the rear end wiggles to the surface to release eggs. At Samoa, these tails appear by the millions on the morning before the last quarter of the moon in October and November. The natives regard them as a great delicacy and gather large quantities, to be wrapped in breadfruit leaves and cooked. In the Florida Keys the worms swarm at the time of the full moon in June, and that is known as the best time to fish for tarpon, which feast on them.

Perhaps the most sophisticated of the moon-regulated sea-dwellers is the grunion, a six-inch member of the prolific silver-side family. It lives only off the California coast, chiefly in the 300 miles from Point Conception, near Santa Barbara, south along Baja California. It spawns during the six months from March to August but only in a three-hour period after high tide on three or four nights after full moon or new moon. California has mixed tides and spawning takes place only after the higher high, which occurs at night during this time of year. The arrival of the grunion to lay eggs in the sand of beaches is the occasion for a festive harvest by people who need only to pick the fish up and toss them into pails.

"The spawning run is heralded by a few lone scouts (usually males) that swim in with a wave," wrote Boyd W. Walker, a University of California biologist at Los Angeles. "Spawning usually starts about twenty minutes after the first fish arrive, reaching a peak of activity about an hour after the start of the run and lasting from thirty minutes to sixty minutes, on the average.

"During a good run, thousands of fish may be flopping on the wet sand at one time, turning it into a vast sheet of shimmering silver."

The female digs her tail into the sand, which is almost fluid as the tide retreats, and males curve around her to discharge their milt as she lays her eggs two or three inches below the surface.

Waves from the lower tides that follow may wash more sand over the eggs until they may be buried at least eight inches deep.

The next spring tide a fortnight later washes the eggs out of the sand. Grunion fry hatch within two or three minutes and are carried out to sea, to grow and, in their turn, respond to the moon's call to the beach. Walker made this further observation:

> Since the waves tend to erode sand from the beach as the tide rises, and deposit sand as the tide falls, it is obvious that if the grunion spawned on a rising tide the eggs would be washed out by succeeding waves. This danger is averted by the fact that spawning is usually confined to the falling tide.
>
> Furthermore, the grunion almost always spawn on a descending tide series, when succeeding tides are lower than those of the previous night — the eggs would be washed out if spawned on the ascending series. There are only three or four nights each month when conditions are right for spawning.

Grunion eggs can be hatched in the laboratory but first they must be agitated, as if being washed by the high tide surf.

Another conspicuous breeder at the time of spring tides is the horseshoe crab; since it has had 200 million years or so to perfect its egg-laying system, it must have found an efficient means of reproduction. The horseshoe crab is really not a crab, but a relative of the spiders and a sort of living fossil, not too different from its ancestors, the trilobites, which flourished about 500 million years ago. It has a large, helmet-shaped, horny shell, which is so thin and tough that it looks as if it were made of dirty brown plastic, and a spike of a tail, useful chiefly for righting itself when turned on its back. Beneath the shell there is a very small body, which appears to consist mostly of gills and six pairs of legs. One thinks of opening the hood of a limousine and finding nothing but a motorcycle engine inside. Perhaps the lack of a substantial body has saved it from the voracious appetites of the deep. It is hardly worth bothering to eat, though gulls nibble when they find it ashore.

Horseshoe crabs spend most of their time crawling over muddy bottoms, and there must be millions of them on the western side of the Atlantic. (The only others in the world belong to a species living on the western shore of the Pacific.)

Full moon in the spring is mating time for horseshoe crabs. I have watched them in Florida in March and in Massachusetts in June. The male is much smaller than the female and he clings to her back as she slowly crawls out of the water. In their eagerness, two males may be attached. The female digs a deep hole in the sand and lays about 10,000 eggs in it, the males sprawl awkwardly to supply sperm, the hole is covered, and the parents make their cumbersome trip back into the water, not to emerge again for another year. When a high tide washes the beach and churning sand particles scratch the blue-green shells of the eggs, the babies emerge, looking like tiny trilobites because they have yet to grow tails. Eggs and young are easy prey for shore birds but when progeny number in the millions the species will surely endure.

It is obvious how—through evolutionary experimentation in spawning and feeding—crabs and oysters, gulls and grunion, could adapt to a lunar cycle in their environment. But the mechanisms by which such creatures time their activities with the tides and seasons are by no means obvious. Gulls, which are smart and are the world's most gluttonous opportunists, presumably can learn through experience and observation where and when to find food. But they appear to be too enterprising merely to wait at a shore for the tide to recede. The Reef Heron in Australia—which lives in rookeries inland, out of sight of the shore—feeds during daytime low tides, arriving at the shore to feed fifty minutes later each day. What kind of a clock does the heron have? And how does a grunion know when spring tide is due?

There are as yet no complete answers to such questions but a great deal has been learned in recent years about the rhythms of living things. The pioneer in the field was Erwin Bünning, a German botanist, who found that the "sleeping" and "waking" of bean leaves (they drooped or lifted) was controlled by some sort of internal clock. And he studied fruit flies, finding that they emerge from their pupal cases at dawn, and continued to do so through fifteen generations, even when kept in constant light and temperature with no clue as to when the sun rose.

Since then many experiments have been conducted with many organisms—sprouting potatoes, woodchucks, honeybees, cockroaches, tobacco, flatworms, human beings, and others—and the existence of a variety of rhythms has been demonstrated. Most of them are related to day and night, to solar and lunar cycles. In some cases they involve such mysteries as the ability of birds to migrate with precision and bees to direct one another to sources of nectar.

I have referred to the activity of fiddler crabs in synchronization with the tides. When Frank A. Brown, Jr., and H. Marguerite Webb made those observations thirty years ago at Woods Hole, they were studying another cycle. Fiddler crabs are dark colored during the day and at night they become pale. The investigators found that the color cycle persisted on a twenty-four-hour timetable, regardless of the temperature and even when the crabs were kept in continuous darkness. The timing of the cycle could be made to shift—so that crabs were dark at night and pale in day, by exposing them to out-of-phase light—but the twenty-four-hour schedule for color change remained.

In addition to the solar cycle, the fiddler crabs had a cycle of activity tuned to the moon. As if they were scuttling around the beach at low tide, this activity peaked fifty minutes later each day as the time of the tide changed.

Brown then had fifteen oysters from the Connecticut shore shipped to his laboratory at Northwestern University in Evanston, Illinois, where he is a professor of biology. He rigged the oysters so that a record could be kept of when they opened their shells, which oysters customarily do at high tide. Lo and behold, the oysters continued to open their shells twice a day on schedule with the high tides back home in Long Island Sound. After two weeks, the oysters changed their rhythm to that of the upper and lower transits of the moon at Chicago—"high tide" time in the Midwest.

Since then, experiments have been performed on whole menageries of organisms with all kinds of exquisite variations and with a variety of resultant discoveries. For example, both the quahog and a marine snail, to whom high tide is important, have their greatest activity at the high tide when the moon is at its zenith, yet less activity at the high tide when the moon is at nadir. The rat, on the other hand, doesn't care at all about the tides but shows minimum activity when the moon is at zenith and peaks when the moon is at nadir. Trying to make sense out of such patterns has complicated the task of biologists, whose one great basic goal is to discover the mechanisms of timing and how they work.

Brown has been the leader of researchers who believe there must be external influences. He is convinced that no matter how carefully controlled a laboratory environment is, some subtle clues get through to an organism, adjusting its clock to particular rhythms. He has found sensitivity to such things as barometric pressure, magnetic and electrical fields, and even cosmic rays, which in turn reflect movements of the sun and moon — yes, and the stars. Eleven years of study of the consumption of oxygen by potatoes showed a daily rhythm of metabolism in *sidereal time* — that is, through the rotation of the earth with respect to the stars.

In an effort to determine whether indeed the rotation of the earth is a factor in rhythms, Karl C. Hamner, University of California botanist, took a collection of hamsters, fruit flies, bean plants, and bread mold to the South Pole and observed their behavior on turntables which were spinning once every twenty-four hours in reverse of the direction of the rotation of the earth: this gave them zero rotation in sidereal space. The fruit flies continued to emerge from their pupal cases according to their regular schedule. Hamsters, beans, and mold also did not change their rhythms.

Such experiments have left investigators with the certainty that organisms have internal clocks that control their rhythms. Brown and scientists of like convictions believe that these clocks are influenced or adjusted by external forces. A force of this kind is referred to as a *Zeitgeber* — time giver.

The layman might jump to the conclusion that a beach-dwelling organism with tidal rhythms merely responds to the advance or retreat of the waters, to getting wet or getting dry. Not so, in the opinion of Palmer, who studied the *Hantzchia*, the alga that hides in the sand at high tide. He writes:

> The phase of the tidal rhythms is, paradoxically, not set by the periodic wetting of high tides. Temperature and pressure changes, delivered by the tides, are the most important phase setters for intertidal organisms whose habitat is not exposed to the direct pounding of the surf. Mechanical agitation is an effective phase setter in organisms of the exposed beach, such as the sand-hopper.

Some scientists have concentrated on trying to find internal clocks, convinced that although an external *Zeitgeber* may "set" a timekeeper, there are mechanisms in cells that operate on a molecular level and can be passed on from generation to generation. Among the leaders is J. Woodland Hastings of Harvard, who has extensively studied *Gonyaulax* polyedra, a single-celled alga which is best known for its blue-green luminescence. You have seen it if you have been on the ocean at night, when there are millions of sparkles in the water as the bow of a boat cuts through the waves. *Gonyaulax* lights up when it is agitated, and this happens even in a laboratory if a test tube containing the algae is shaken.

Studies of *Gonyaulax* in the lab showed that when it was kept in constant illumination, either in darkness or in light, it would luminesce only during the period of astronomical night-time. Furthermore, it has other daily rhythms; one is related to photosynthesis (converting sunlight into chemical energy) which, as might be expected, is active during the daytime period. And most of the cell division (the process by which it reproduces) occurs during a five-hour period that spans what in the open ocean would be dawn. The alga loses its rhythmicity when chilled and exposed to a bright light (both are required); when returned to normal conditions in the laboratory, it has a different phase, determined by the time of return.

Whatever the mechanisms, they are complex. In a report on discussions by biologists, Hastings wrote:

> Theorists of circadian rhythmicity, unafraid to draw facts from a variety of phyla and cell types, have had little trouble assembling models according to which the universal circadian mechanism is whatever is currently in vogue biochemically: first soluble enzyme kinetics, then nuclear message transcription, then membranes and the second messenger. Could it be that all these aspects of cell chemistry are susceptible to circadian modification of their regulatory dynamics, and that the clock, volatile as a ghost, lurks now in one room, now in another, in different cell types?

Whatever the chemistry, the basic rhythm may be that of the division of cells. Specific times may be required by DNA, the master molecule, to replicate itself, just as a specific time is required to play a piece of music from a tape.

The different research approaches of Brown and Hastings to the problem of biological clocks may lead to the same end. During their summers at the Marine Biological Laboratory at Woods Hole, they are neighbors on a secluded and forested lane and they have ample occasion for the exchange of ideas.

Brown continues to have confidence in the importance of external controls of rhythms. "There is no proof that there is an independent, internal clock in organisms," he says. "I don't think of an organism as being like a clock with a pendulum that controls its time keeping. I think of it rather as comparable to an electric clock, which is timed by the sixty-cycle current. When it is plugged in to a sixty-cycle line, it keeps perfect time but when it is not, it doesn't."

In recent research, as yet unpublished, Brown has found evidence that bean seeds can affect other beans nearby, and he believes that the means of communication is by very subtle oscillating magnetic waves. This belief begins to approach an almost mystical area in which most scientists don't want to be caught — an area in which all living things can be thought of as

198

being tuned to the rhythms of nature in ways we cannot comprehend.

"There is so much that we have yet to learn!" Dr. Brown said recently. One recent development of interest, he feels, is the finding by a Swiss scientist, Fritz Schneider, that beetles are able to sense minute changes in mass-space distribution — in other words, that they can sense gravitational changes.

In a book that appeared in 1978, *The Lunar Effect,* Arnold L. Lieber speculated that "we shall discover there are 'gravoreceptors' within the human body" which sense tidal forces. A Miami psychiatrist, Lieber examined a wide range of phenomena which he sought to explain by the moon's influence. When Fergus Wood alerted the nation to extreme proxigean tides on January 8 and February 7, 1974 (see Chapter 5), Lieber alerted police and medical officials in Dade County, Florida, predicting "a general disturbance in human behavior."

"All hell broke loose," Lieber observes. Murders during the first three weeks of the year were triple the number for the entire month of January in 1973, and some of them were notably bizarre and brutal. Psychiatric emergency room visits during the month increased by 35 percent over the previous month. After examining these statistics, and those from a number of other studies, Lieber came up with a "biological tides hypothesis." He said:

> Because the body is composed of 80 percent water and 20 percent "land," or solids, it is reasonable to assume that gravity exerts a direct effect on the water mass of the body, just as it does on the water mass of the planet.

And there is also an indirect electromagnetic effect on the nervous system, he thinks. (I mentioned magnetic tides in Chapter 8.) But if the moon's gravitational force is only 1/10,000,000th of the earth's, as I pointed out in Chapter 1, it stretches the imagination to suppose that body fluids respond to the moon as the oceans do. And to conclude that there are also electromagnetic effects requires some fancy extrapolation.

Still, there is no doubt that rhythms do exist and some of them, even in humans, seem to be synchronized with the moon. A significant example is provided by the case of a young blind man who was studied by the Sleep Research Center at Stanford University School of Medicine.

The subject was a postgraduate student who had been blind from birth. For several years he had suffered insomnia at night and sleepiness during the day. He was placed in a hospital for a month and encouraged to sleep whenever he wished. Continuous testing of body temperature, alertness, performance, cortisol secretion, and urinary electrolyte excretion showed a daily rhythm that could not be distinguished from the lunar day of 24.84 hours. He was then put on a strictly 24-hour schedule and required to stay in bed from 11:00 P.M. to 7:00 A.M. The troublesome symptoms returned.

"Furthermore," reported the researchers, "throughout the ad-lib sleep study, there was a remarkable coincidence between his sleep onset and the local low tide." The report also noted that "Our subsequent survey of 50 subjects with varying degrees of blindness revealed that 38 complained of a significant sleep-wake disorder."

The equivalence of the female menstrual period and the monthly cycle of the moon has not only been recognized by folk wisdom, probably for thousands of years, but is an intriguing problem for medical science today. Actually, although twenty-eight days are generally thought of as the length of the menstrual period, studies have shown that the average period is closer to 29.53 days, the *synodic cycle*, from new moon to new moon. Can this be a mere coincidence? Certainly the Latin roots of *mensis* for month and *menstruus* for monthly, of *mensura* for measuring; the Old English *mona* for moon and *monath* for month, have identified the menses and menstruation (as well as mensuration) in our language with the measuring of time by the monthly circuit of the moon.

Charles Darwin thought that the menstrual cycle might be a vestige of physiological rhythms inherited from our sea-dwelling ancestors. An objection to this idea is that menstruation does not occur among our mammal kin, except in certain monkeys and the apes; and their estrus is generally seasonal. But some primates menstruate at new moon and ovulate at full moon and the sexual activity of certain ones are synchronized with the full moon.

Walter and Abraham Menaker studied 250,000 live births in New York and found that there were more during the half of the lunar cycle beginning just before full moon than during the half-cycle beginning just before new moon. Since the estimated mean duration of pregnancy is 266 days (compared to 265.8 days which is the total of nine synodic lunar months) they concluded that "The peak of conception and probably ovulation appears to occur at full moon or a day before it." Walter Menaker later studied a half-million births and found the highest birthrate to be during the brightest half of the lunar cycle, centered on full moon.

On the theory that the light of the moon might influence female rhythms, Edmond M. Dewan, as an associate at Brandeis University, with Miriam F. Menkin and John Rock conducted experiments with artificial light. Dewan explained:

> The hypothesis that ovulation occurred in the human at full moon during evolutionary times has an interesting implication. Natural selection would favor the tendency to mate at this time of the month. Contrariwise, a tendency to mate at this time of the month (say for example due to the ability to see at night) would favor, again by natural selection, the tendency to ovulate at the time of full moon. If this speculation were correct, it would provide not only an evolutionary explanation of the phenomenon here studied. It would also explain, on a rational basis, the cause of the well known "romantic" effect of the full moon.

In one study, sixteen women with a history of irregular or abnormally long menstrual cycles were exposed to all-night light beginning with the fourteenth day of their cycles — with the hope that (simulating the moon's light) it would trigger ovulation on the fifteenth day and result in menstruation at the end of twenty-nine days. The procedure was simple: a table lamp would be placed on the floor, ten feet from the patient's bed, and left shining all night for four nights. The same women were also studied during control cycles when no lights were used.

During the control periods there was a good deal of variation in the length of cycles—from 25 to 38 days. But during the experimental periods, when the lights burned all night, the greatest number of women had cycles of about 29 days (though there was still variation). The "full moon effect" was so interesting that Dewan hopes the experiments can be conducted with a larger sampling of subjects.

Perhaps the moon has had an even more profound influence on the human race than any exerted through moonlight. Nearly two decades ago, Sir Alister Hardy, professor of zoology and comparative anatomy at Oxford University, propounded a bold speculation: that our tool-using primate ancestor was not a predatory hunter in the forests or plains but was a creature of the sea. The human body is not fur bearing, for protection from cold or sun, but streamlined as if for swimming, he observed. And if a sea otter can float on its back and crack open sea urchins with a rock, so could primates learn to use such tools. Early man may thus have been subject to tidal rhythms.

So far as I know, no evidence has ever been found to support such a theory and perhaps none can be. But Elaine Morgan, an Englishwoman, carried the idea further in a book, *The Descent of Woman*. With a fervent feminist approach, she speculated that the true parent of the human race was a mother, not a father. She contended that anatomically the female is constructed for copulating under water and suckling under water and that it would have been a woman who used the first tools, perhaps by cracking open oysters with rocks. All of this would have put her in tune with the tides and the moon.

At least we are sure that *Femina sapiens* did not emulate the gulls and learn to crack shellfish by flying into the air and dropping them on a rocky beach.

NOTES

"Gulls" by E. A. Muir, copyright © 1965 by *Harper's Magazine* is reprinted by permission of the author.

Among the sources for references to marine life are: *Under the Sea Wind* by Rachel L. Carson (Oxford University Press, London, 1941, New American Library, New York, 1955); *The Edge of the Sea* by Carson (Houghton Mifflin, 1955); *The Infinite River* by William Amos (Random House, New York, 1970); *The Wonderful World of the Seashore* by Albro Gaul (Appleton-Century-Crofts, New York, 1955); *The Sea-Beach At Ebb-Tide* by Augusta Foote Arnold (Century Company, 1901, Dover, New York, 1968); *At the Turn of the Tide* by Richard Perry (Taplinger, New York, 1972).

Hantzchia virgata and other organisms were discussed by John D. Palmer in "Biological Clocks of the Tidal Zone" in *Scientific American* (Vol. 232, No. 2, February 1975) and by Boyd W. Walker in "The Timely Grunion" in *Natural History* (Vol. LXVIII, No. 6, July 1959). Other sources on rhythms included *An Introduction to Biological Rhythms* by Palmer, with contributions by Frank A. Brown, Jr., and Leland N. Edmunds, Jr. (Academic Press, New York, 1976); *The Biological Clock* by Brown, Palmer, and J. Woodland Hastings (Academic Press, New York, 1970); *The Physiological Clock* by Erwin Bünning (Springer-Verlag, New York, 1964); *The Living Clocks* by Ritchie R. Ward (Alfred A. Knopf, New York, 1971); "Conditionality of Circadian Rhythmicity: Synergistic Action of Light and Temperature" by Hastings, David Njus, and Laura McMurry (*Journal of Comparative Physiology*, Vol. 117, pp. 335-344, 1977); "The Molecular Basis of Circadian Rhythms," report by Hastings, Hans-Georg Schweiger, and others on the Dahlem Workshop (Life Sciences Research Report 1); "Blind Man Living in Normal Society Has Circadian Rhythms of 24.9 Hours" by L. E. Miles, D. M. Raynal, and M. A. Wilson (*Science*, October 28, 1977, Vol. 198, No. 4315, pp. 421-423). Stephen Jay Gould's articles in *Natural History* on evolution were informative and his personal guidance at Harvard and that of Dr. Brown at Woods Hole encouraging.

Sources on menstrual cycles were *The Curse, A Cultural History of Menstruation* by Janice Delaney, Mary Jane Lupton and Emily Toth (E. P. Dutton, New York, 1976); *Body Time* by Gay Gaer Luce (Pantheon, New York, 1971); "Lunar Periodicity in Human Reproduction: A Likely Unit of Biological Time" by Walter and Abraham Menaker (*American Journal of Obstetrical Gynecology*, Vol. 77, pp. 905-914, 1959); "Lunar Periodicity With Reference to Live Births" by Walter Menaker (AJOG, Vol. 98, pp. 1002-1004, 1967); "Effect of Photic Stimulation on the Human Menstrual Cycle" by Edmond M. Dewan, Miriam F. Menkin, and John Rock (*Photochemistry and Photobiology*, Vol. 27, pp. 581-585, 1978).

Sir Alister Hardy's article, "Was Man More Aquatic in the Past?" appeared in *New Scientist* (Vol. 7, pp. 642-645, 1960) and *The Descent of Woman* by Elaine Morgan was published by Stein & Day, New York, in 1972.

The Lunar Effect by Arnold L. Lieber (Doubleday/Anchor Press, Garden City, N.J., 1978) offers adventurous conclusions on tide-like effects and periods in living things.

Tides of War

In addition to satisfying tidal and sunlight req-
uisites, the planners, like Bottom and Quince,
had to 'look in the almanack; find out moonshine, find out moonshine' of sufficient intensity
to please the aviators; and they managed to provide them with a full moon.   SAMUEL ELIOT
MORISON

COMMANDER D. H. Mac-
millan, an English authority on the tides, has said:

It is interesting to speculate upon the path history
would have taken if the British Islands had been
surrounded by a tideless sea. The segregation of
hardy peoples in migration to our islands for centuries and the acquiring of distinctive qualities in
the mastery of tidal phenomena, constituting a
world force in sea power, would not have come,
and the whole of the West would probably have
been assimilated to Mediterranean civilization
without any insuperable natural barriers apart
from bad weather. . . .

It is, therefore, reasonable to assume that the
influence of the tides on world history is not in-
considerable, and that a fortuitous combination
of large tidal ranges and rich mineral resources in

204

the British Islands has, despite changing aspects,
determined the character, course, and ascen-
dancy of the Anglo-Saxon peoples and the expansion of their free institutions and cultures.

Background music for such reflections obviously should be "Hail
Britannia . . . Britannia rules the waves!" In the migrations of
which MacMillan writes, the invading Angles, Saxons, Normans,
and others were hardy enough not to be turned back by the tides;
and the greatest of the Mediterraneans, Julius Caesar, managed to
surmount them, though indeed his first try was almost disastrous.

It was late in the summer of 55 B.C. when Caesar, after busy
months of slaying Germans, taking hostages, burning villages, and
crossing the Rhine, decided to extend Rome's benevolent protection to the Britons. He could learn little about Britain from traders
and sent a spy to investigate the island. The scout returned after
five days, during which he did not dare land, with little information.

Caesar had assembled eighty transports to carry two legions—
7000 or more men—plus eighteen boats for the cavalry. The fleet
set off from what is now Boulogne shortly after midnight on August
25. Arriving near Dover about 9 A.M., Caesar found the white cliffs
lined with barbarians prepared to hurl rocks down on landing
troops, so with the turn of the tide he led the fleet about seven miles
up the Channel to where there were no cliffs. But the invaders
still had the problem that would plague many other amphibious
forces. Caesar wrote:

The ships, on account of their size, could
not run ashore, except in deep water; the
troops—though they did not know the ground,
were loaded with the great and grievous weight
of their arms—had nevertheless at one and the
same time to leap down from the vessels, to
stand firm in the waves, and to fight the enemy.

Caesar sent his war galleys, with slings, arrows, and artillery, to
attack the defenders' flank and as soon as the troops could gain
dry land they routed the Britons. Meanwhile the transports with
cavalry had been delayed; when they finally arrived off the

British coast four days later, a violent storm drove them back to the continent.

"That same night, as it chanced, the moon was full, the day of the month which usually makes the highest tides in the ocean, a fact unknown to our men," Caesar related. Actually the highest tide would follow the moon by a day and a half; but the series of spring tides would have begun and, with a tempest raging, no doubt the tide was quite high. It seems possible, however, that Caesar was building an alibi in his report to the Senators back home.

In any case, the Romans had drawn their war galleys up on the beach, thinking them safe, while the transports were at anchor. The high tide filled the beached boats, and surf gave the whole fleet a buffeting. "Several ships went to pieces," Caesar wrote, "and the others, by loss of cordage, anchors and the rest of their tackle, were rendered useless for sailing."

The invaders did not have a large supply of corn and, if stranded, were in trouble. The Britons, who by this time had been more or less subdued, realized this and renewed the fighting. But ships were repaired with timber and bronze fittings from the damaged ones and Caesar was able to evacuate his troops to the continent.

Before making his annual winter trip to Rome, Caesar gave orders for the construction of an invasion fleet for the next year. The transports were to be broader and of shallow draft for beaching, and were to be fitted with oars as well as sails.

By the summer of 54 B.C., 600 such boats, plus 28 men-of-war, had been built. The fleet left port about July 6, ran up the Channel with the tide, and landed north of Dover on a sandy beach. This time Caesar had five legions and 2000 cavalry and they found no defenders on the shore. The Britons had assembled to resist the landing, he learned later, but upon seeing the size of the fleet they had retreated.

While Caesar and his troops were pursuing the Britons, a violent storm occurred, as at the time of the first invasion (but Caesar wrote no account of high tides this time). Some forty vessels were lost. Caesar had all the ships beached and spent the next ten days getting repairs made on damaged ones. Then he proceeded to vanquish the natives who, like the American Indians and Viet Cong many centuries later, made it difficult by melting away into the woods and then dashing out to attack advancing columns. By September, however, Caesar dominated south-eastern Britain and, with many prisoners and hostages, sailed back to Gaul before the equinoctial storms began.

There were many conquests and landings in the next 2,000 years and some failed schemes. Napoleon and Hitler assembled fleets with the determination to repeat Caesar's invasions of Britain. One proposal considered by the Nazis was to build "war crocodiles," huge concrete barges—each of which would carry 200 men or several tanks up to a beach, meanwhile serving as seagoing forts. But by September 21, 1940, one date set by Hitler for the invasion, British bombers had destroyed 21 transports and 214 barges—twelve percent of the fleet assembled by the Germans. The cross-Channel operation, named Sea Lion, was called off. Elaborate technological and tactical development by the Allies would be necessary to make large-scale amphibious operations successful, including the greatest one of all times by a fleet moving from Britain to Gaul.

Not having made a landing since 1898, the American military was rather indifferent to the problem until 1924, when the Marines were sent into a practice assault at Culebra Island near Puerto Rico. Ten more years passed before they executed a comprehensive fleet landing. Such exercises were continued, however mechanically. A manual for joint Army and Navy operations was issued, and development of landing craft began in the thirties. So the United States was not totally unprepared for the challenges that were immediately apparent after Pearl Harbor. Amphibious operations are extremely complicated, however, and neither trained manpower nor equipment was adequate.

The first real test came in the landing on North African beaches in November 1942. The Marines were busy in the Pacific, where they had made relatively easy landings on Guadalcanal and other Solomon Islands (the most desperate fighting came later) in August. Establishment of a beachhead in Morocco therefore had to be an Army–Navy operation.

An armada of 102 ships took part in the invasion called Torch, the objective of which was to land 35,000 men and 250 tanks at three points and begin the task of driving the Germans out of North Africa. Three main kinds of landing craft were used: the plywood "Higgins boat," which had a square bow, and two boats with ramps, one of which would carry a truck. No detailed charts of the shore were available and little was known about it. For maximum surprise, the forces needed a night landing, which was planned to start at 4:00 A.M. on November 8—a time when the moon would be in its last quarter and the tide would be ebbing. On the day before the landing, meteorologists predicted a fifteen-foot surf, and there was doubt that a landing could be made; but strategists decided to take a gamble, and the weather moderated.

Low tide on the beach would be at 7:44 A.M., less than an hour after sunrise, and a critical time in the assault. (The commander, Rear Admiral H. Kent Hewitt, had asked that the landings be postponed for a week to take advantage of a rising tide but a decision was made against such a change.) As the assault began the process of getting the boats away from the ships was delayed, so the start turned out to be at 5:00 A.M., still closer to low water.

Of 25 boats from the Transport Charles Carroll, 18 were wrecked on the first landing and 5 more after returning with a second load. Of 32 boats from the Transport Leonard Wood, 21 were wrecked on the first landing and 8 more in the hours that followed. Of the total of 629 boats used, 216 were damaged or destroyed.

"Inexperience, darkness and the necessity of landing on a falling tide were the principal causes of the loss of boats," wrote Samuel Eliot Morison who was with the expedition as the Navy's historian.

> Salvage facilities were inadequate, and if help was not forthcoming when the tide came in and a stranded boat floated, it broached-to and rolled over in the surf. Even more damage was done by boats crowding together on the beaches, so that waves banged them one against the other. Stranded boats became targets for enemy gunners and a number were lost on rocky ledges.

But troops were delivered, the surprise was complete, and an hour passed before defenders' guns began firing.

This was the essence of the story of the landing at Fedhala, the beach closest to Casablanca. Landings at the two other points, Port Lyautey and Safi, were also successful. And although many ships would be sunk and many lives lost, by June of 1943 Algiers and Tunisia as well as Morocco had been liberated and the way had been opened for landings in Sicily and Italy.

Meanwhile the Navy and the Marines had been desperately engaged in wresting control of strategic islands in the Pacific from the Japanese. That campaign led, just eleven days after the landings in North Africa, to Tarawa, an atoll in the Gilbert Islands. An airfield on the heavily fortified island of Betio was the objective. The coral reefs were natural hazards that added to the defenses; at low tide one could walk between the twenty-five islands of the atoll. Admiral Morison wrote:

> No accurate tide tables for the Gilberts existed. They were unpredictable. No one could foretell whether there would occur on 20 November a so-called "dodging" tide. That is an irregular neap tide which ebbs and flows several times a day at unpredictable intervals, and maintains a constant level for hours on end. For instance, the dodging tide that occurred in the following lunar month rose to a level 4 feet above the outer edge of the reef and stood still for 3 hours, fell 1 foot in the next 3 hours, stood still 2½ hours more, fell 2 feet by 1815 and gradually rose to a 3.4 foot level at 2000, stood still 2 hours, and so on. That was a "high dodging" tide which would have helped the Marines by keeping a sufficient depth of water over the reef to float a landing craft. But on 20 November they had the bad luck to draw a "low dodging" tide that would admit no boats over the reef. It was impossible to predict whether there would be a low or a high one or an ordinary neap tide, which would have been all right for at least six hours on the 20th.

The moon was in the last quarter. Spring tides, due on November 22, would cover the beach up to the Japanese barricades. For this and other reasons, such as the danger of a wind that would make landing impossible, Rear Admiral Richmond Kelley Turner considered the two-to-one chance of a favorable tide on the 20th, "took the long chance, and lost."

Even with a friendly tide the Marines would have undergone a cruel episode, for there were 4500 Japanese troops on Betio, an island less than two miles long and 500 yards across, and they had constructed blastproof shelters of concrete, coconut logs, and sand. Even after an aerial bombing followed by a two-and-a-half-hour bombardment by three battleships, four cruisers, and a number of destroyers, the enemy was able to direct a murderous fire at the landing force.

Boats had to come more than six miles from the ships. The landings began at 9:00 A.M. and immediately it was found that the Higgins boats couldn't pass over the coral reef. This meant that troops had to be shuttled in amphtracs — 25-foot amphibious tractors which could crawl over the coral but were duck-like targets for Japanese gunners. Of 125 that were used, 90 were destroyed.

The distinguished correspondent Robert Sherrod was with a group of Marines who were transferred from a Higgins boat to an amphtrac, then discharged into neck-deep water. They had to walk 700 yards into the face of machine gun fire before they could take shelter at the base of a sea wall of coconut logs. On every side men were being killed, but by nightfall 3000 had been landed.

The second day was almost as bad, but a "normal" tide came in at noon, enabling light tanks to be landed. The high water threatened to drown wounded men lying on the reef before they could be rescued. By the end of the day Japanese strength was ebbing, though two more days were required to win full control of the island. Of 18,313 Americans engaged in the battle, 1,009 lost their lives and 2101 were wounded. Of the defenders, only one officer, 16 enlisted men, and 129 Korean workers were taken prisoner. The rest fought until they were killed.

When high tide brought three feet of water up to the sea wall where the first assault troops had taken shelter, Sherrod realized that though the scanty tide had hampered the landing, without it "the Marines would have had no beachhead at all!"

Amphibious forces learned many things at Tarawa and although there would be agony in other landings, the lessons were put to good use. While Pacific forces struggled to capture the stepping stones toward Japan, planning went forward for the greatest invasion in history—the return of Allied armies to France.

Tidal predictions were crucial in the timing of the landing, the immediate purpose of which was to cut off the Cherbourg Peninsula and gain ports—in preparation for the hard drive across France, Belgium, and Holland to Germany. Probably no tides of the world have been more closely studied than those of the English Channel, and the British were ready with detailed data on the Norman coast, where the average range is eighteen feet and the highest tide, twenty-five feet.

Germans had lined the beaches with concrete and steel obstacles to prevent a landing. The Navy wanted to make the assault at low tide so that the obstacles could be seen and destroyed by demolition crews from boats in the water beyond the thorny barricade. The Army wanted to land close to high tide so that troops would have narrower strips of exposed beaches to cross; it also wanted a second high tide during the same day to land reinforcements and materiel. The air forces wanted a full moon on the night before the landing to facilitate bombing, and then forty minutes of daylight to complete this preliminary pounding of the defenders. The invasion force needed to cross the Channel under the cover of darkness.

June 5, 6, and 7 were found to provide the proper moon and tide conditions but the Army and Navy would have to compromise on the landing time: they decided on a landing from one to three hours after low water. Sunrise would be at 5:58 (double daylight saving time). Five different H-hours, from 6:30 to 7:55, were set for the five beaches that would be used.

Weather was the imponderable. If it proved stormy, the invasion would have to wait until the next spring tides, two weeks or a month later.

"Consequences would ensue that were almost terrifying to contemplate," wrote Dwight D. Eisenhower on the possibility of canceling (in *Crusade in Europe*).

> Secrecy would be lost. Assault troops would be unloaded and crowded back into assembly areas enclosed in barbed wire, where their original places would already have been taken by those to follow in subsequent waves. Complicated movement tables would be scrapped. Morale would drop. A wait of at least fourteen days, possibly twenty-eight, would be necessary—a sort of suspended animation involving more than 2,000,000 men. The good-weather period available for major campaigning would become still shorter and the enemy's defenses would become still stronger! The whole of the United Kingdom would become quickly aware that something had gone wrong and national discouragement there and in America could lead to unforeseen results. Finally, always lurking in the background was the knowledge that the enemy was developing new, and presumably effective, secret weapons on the French coast. What the effect of these would be on our crowded harbors, especially at Plymouth and Portsmouth, we could not even guess.
>
> It was a tense period, made even worse by the fact that the one thing that could give us this disastrous setback was entirely outside our control . . . . If really bad weather should endure permanently, the Nazi would need nothing else to defend the Normandy coast!

When General Eisenhower and other leaders met at 4 A.M. on June 4 to make a final decision on whether to launch the invasion next day, meteorologists told them that low clouds, high winds, and high waves would make a landing extremely hazardous. The assault was postponed, though some ships from northern ports were already at sea and would have to return and refuel. "Tension mounted even higher," Eisenhower wrote, "because the inescapable consequences of postponement were almost too bitter to contemplate."

But when the leaders met again the next morning, meteorologists promised thirty-six hours of relatively good weather beginning on June 6, though this period would be followed by bad weather which would make further operations difficult. Eisenhower decided the consequences of delay justified the risk.

The assaulting force was almost incredibly large and the details of coordination and supply almost beyond comprehension. Ships and landing craft totaled 9000 and were protected by 702 warships and 25 minesweeper flotillas. They carried 20,111 vehicles and 176,475 men. Airborne troops were landed inland by 1384 transport planes and 876 gliders. A total of 2219 bombers prepared the way—not very effectively because of cloudy weather.

The attack front extended for sixty miles, with the Americans landing at Utah Beach on the peninsula, aiming for the capture of Cherbourg; at Omaha Beach, farther to the east. British and Canadians in the Gold, Juno, and Sword forces sought to seize the shore beyond, as far as the Orne River.

The landing on Utah Beach surprisingly did not encounter enemy fire, and there was no surf but the tide was rising so rapidly that men had to wade through a hundred yards of water. One LCI (Landing Craft, Infantry) commander found that he was delivering soldiers at a point where water was over their heads, and he had to retract and beach five times. The assault force lost 197 men—60 of them in the water.

Omaha Beach was a different matter. A strong tidal current threw landing craft off course, and there was heavy surf. Demolition crews opened five channels and three other partial ones through the obstacles, but the tide was rising as much as a foot in eight minutes and they could not complete their work. Enemy gunfire from the high bluffs was intense and casualties totaled 3000. With heavy packs, men wounded while floundering in the water were not likely to escape drowning. Fifty landing craft were hit or sunk and only 43 of 96 tanks reached the beach.

Nevertheless, by the end of the day, Omaha was secured. The British and Canadians had also established their beachheads.

On June 19 there was a violent storm with very high tides—the worst June storm in 40 years—which wrecked more than 300 vessels and damaged beyond repair a prefabricated landing dock called "the mulberry", which had been brought from England. However, American troops took Cherbourg in twenty days after the landing and the armies drove eastward. Paris was surrendered on August 25 and all of France was soon regained.

No one who took part in a World War II amphibious landing could have wanted, or expected, to participate in another. But just five years after the war's end, Americans again faced the need for a beachhead — this time in the United Nations response to the Communist invasion of South Korea on June 25, 1950. Within a week, North Koreans captured Seoul, the capital city, and its port, Inchon, fell soon after.

General Douglas MacArthur, supreme commander of U.N. forces, visualized a landing at Inchon as a necessity and almost immediately set a task force to work on plans, but the proposal was shelved. Finally, in late August, a decision was made to attack Inchon on September 15, only twenty-three days away. Preparations were made frantically.

The date was important because it would be one of maximum tides, which were an absolute necessity. To delay until spring tides in October or November would shorten the time for follow-up Army operations before winter.

Inchon was almost the worst conceivable location for an amphibious landing. The one thing in its favor was that the enemy could not expect the Allies to be so foolhardy as to attempt it. General MacArthur told other military leaders, "I realize that Inchon is a 5000-to-1 gamble, but I'll accept it. I am used to taking those odds. We shall land at Inchon, and I shall crush them!"

The average range of tides at Inchon is 29 feet, with a record of 36 feet — the highest in the Orient. The harbor is approached by a narrow, winding channel with currents of from three to five knots. At low water the harbor has great expanses of mudbanks. Landings would be limited to a four-mile waterfront right in the city, which was lined with sea walls and piers but with little dock space. The U.N. standby ships would have to remain thirty miles away, beyond clusters of islands. The harbor was guarded by a fortified island, Wolmi-do. Presumably the channel would be

Four landing craft are stuck in the mud at low tide, the day after the landing at Inchon, Korea, on September 14, 1950. In spite of the obstacles posed by Inchon's tidal range, the landing was a brilliant success. *(National Archives)*

mined. There was no information on how much dredging had been done. But if Inchon could be captured, so could Seoul, and the Communists' supply lines would be cut.

The assault would have been even more hazardous had it not been for the exploits of a Navy officer, Lieut. Eugene Franklin Clark, who with two interpreters was landed secretly on an island outside the harbor two weeks before the attack. He organized the islanders as an intelligence brigade, radioed information about harbor defenses, and destroyed small enemy forces on other islands. He found that American tide tables were inaccurate and that Japanese tables should be used. He even penetrated the harbor one night at low tide to check on the state of mud flats and, sinking to his waist in the muck, was convinced that a landing at low tide over that wasteland would be impossible. At midnight before the landing began, he went to a lighthouse on a small island, which had been decommissioned by the Reds, and lit the lamp as a signal to the arriving fleet.

The first task was to knock out Wolmi-do. On September 13 six destroyers steamed up the channel at low tide to begin bombarding the fortress at 1:00 P.M. Low tide was chosen partly because it might enable attackers to locate mines that later would be hidden in muddy water, and indeed some were found and were exploded with gunfire. It would also enable the destroyers to fire their guns point-blank at targets. More important, since there was no room to maneuver, the destroyers would have to drop anchor in the channel because of the swift current; it would be better to anchor with their bows facing the incoming tide so that they could make a fast getaway if necessary.

As the destroyers moved into position, planes from carriers began bombing Wolmi-do. One of the destroyers' functions was to draw fire so that hidden enemy batteries could be spotted. The destroyers were among the older ships of the fleet and were regarded as expendable. They did indeed draw fire but only one destroyer was substantially damaged and one man was killed. The next day, while the damaged ship hunted mines, the other five destroyers returned and fired 1732 shells at Wolmi-do and Inchon. Meanwhile, four cruisers were also firing from a distance.

Under the cover of a moonless night and with the tide rising, the destroyers came back a third time, this time escorting troop and assault ships. A third bombardment of Wolmi-do started at 5:45 A.M. and the Marines began landing on the island at 6:31. By 8:07 the fortress had been subdued, with 108 enemy dead, 136 captured and 17 Allied casualties.

From Wolmi-do, tanks and troops would be able to move over a causeway into Inchon while other landings were made at two beaches — the beaches being fifteen-foot sea walls with North Korean trenches and machine gun nests behind them. In preparation for going over the walls, the Allies had built ladders with steel hooks at the top, an ancient device for surmounting walls.

The next high tide would be at 7:19 P.M., thirty-five minutes after sunset. There were just four hours in which the water would be high enough for landing craft. The first wave would hit Red Beach, at the front of the city, and every two minutes six landing craft were to discharge troops. LSTs (Landing Ship, Tanks) would arrive just before high tide; by the time they unloaded tanks, bulldozers, and other vehicles, they would be stranded in the mud. Much of the operation would be at twilight and the battle would undoubtedly be raging all night.

After a forty-five-minute barrage of shells and rockets, the assault began but it did not eliminate Communist defenders or breach many places in the sea wall. The hooks on the ladders were not broad enough to grab the wall in some places and Marines had to be boosted over by their comrades, to be confronted immediately by machine gun and mortar fire.

Marguerite Higgins was one of the correspondents on the fifth wave and took cover with Marines in a notch in the wall. She wrote:

> The sun began to set as we lay there. The yellow glow that it cast over the green-clad Marines produced a technicolor splendor that Hollywood could not have matched. In fact, the strange sunset, combined with the crimson haze of flaming docks, was so spectacular that a movie audience would have considered it overdone.

> Suddenly there was a great surge of water. A huge LST was bearing down on us, its plank door halfway down. A few more feet and we would be smashed. Everyone started shouting and, tracer bullets or no, we got out of there. Two Marines in the back were caught and their feet badly crushed before they could be yanked to safety.

Eight LSTs landed on the beach, which was only a thousand feet wide. To bring them in so soon after the troop landing was an unorthodox tactic because, with thin skins and cargoes of gasoline and ammunition, they were exceedingly vulnerable. When they were stranded at low tide they would not be able to pump sea water to fight fires. But the only way to land the 3000 tons of supplies needed that night was to bring them in at high tide. And they did their job.

Within a day after the landing the Marines had fought their way six miles beyond Inchon, and in ten days Seoul had been captured. There would be a long, hard trail to the end of the Korean War but in one of the most brilliant operations in American history the Navy had outwitted the tides.

NOTES

The quotation by Samuel Eliot Morison is from his epic *History of the United States Naval Operations in World War II*, Vol. XI, *The Invasion of France and Germany* (Little, Brown, Boston, 1957), which also was a principal source of information on the amphibious landing in Normandy. In Vol. II, *Operations in North African Waters* (1947) there is an authoritative account of the landing at Casablanca, and in Vol VII, *Aleutians, Gilberts and Marshalls* (1951), of the landing at Tarawa. I have quoted from *Crusade in Europe* by Dwight D. Eisenhower (Doubleday, Garden City, 1948). Other sources included *Tarawa* by Robert Sherrod (Duell, Sloan & Pearce, New York, 1944).

The War in Korea, Battle Report, Vol. 6, by Capt. Walter Karig, Commander Malcolm W. Cagle, and Lt. Cmdr. Frank A. Mansion (Rhinehart, New York, 1952) gives abundant details on the landing at Inchon. I have quoted from *War in Korea* by Marguerite Higgins (Doubleday, 1951).

Julius Caesar's account of his amphibious landings is given in *The Gallic War*, translated by H. J. Edwards (The Loeb Classical Library, Harvard University Press, 1970). The quotation from Commander D. H. MacMillan appears in his book *Tides*, previously cited.

Tidal Power

From time immemorial man has sensed magic in
the movements of the tides and moon, and, no
doubt, for almost as long, has yearned to become
master of the magic force. It is not surprising that
tidal energy has been put to human use for at
least eight centuries. What is surprising is that no
large-scale effort has been made, until very recently, to harness the vast pulsing energy of the
tides.   THE BAY OF FUNDY TIDAL POWER REVIEW
BOARD.

           A UTOPIAN SOURCE of
energy would be inexhaustible, perpetually dependable, and
nonpolluting. Tidal power meets those specifications. There will
be just as much available in the year 2080 as in 1980. It produces
no smoke, no radioactive wastes, nor even wasted heat to corrupt
the atmosphere and the waters.

No wonder that in an energy-hungry world, where fuel short-
ages threaten dire crises, tidal power projects have received
increased attention. Unfortunately tidal power does have disad-
vantages, the greatest of which is that it is not available twenty-
four hours a day. In its simplest form, it is generated only at certain
stages of the tides, when water flows from a higher level to a

lower one, and that may not be when consumers want the power.

This did not matter in the leisurely times after the eleventh century, when the first tide mills operated in Great Britain, France, and Spain. A bag of grain could wait to be ground when power was available; and if high tide came at noon and midnight, there were no union rules to prevent the miller from working at such awkward hours.

A typical water wheel output was one horsepower-hour to grind one bushel of grain. The tidal power system might not be as desirable as that of a mill on a swift-running stream inland or of a windmill, which is more efficient (when the wind is blowing); but for a coastal community it was adequate. A castle with a moat fed by tides could have a fortified mill that would grind its grain even in time of siege. Such was the situation at Southampton, England, where there was a mill in the God's House Tower, a thirteenth-century barbican which still stands.

The first tide mill in North America was built at Port Royal, now Annapolis, Nova Scotia, in 1607. (The mean tide range there is 22.6 feet.) It was only partially powered by tidal energy and a replica today is powered by electricity. According to Peveril Meigs, an authority on tidal mills, the first tide mill in what is now the United States was a sawmill built in 1634 in the settlement of Sir Ferdinand Gorges at Agamenticus — now York, Maine. There probably was one in Boston by 1631, however, and several were built in surrounding towns in the decades after that.

The first ambitious tidal project in Boston was undertaken in 1634, when a cove and marshland on the north side were granted to a group of entrepreneurs. A dam was built over an old Indian trail across the marshes to enclose the cove for a mill pond, with a ten-foot gate to admit the tide. Two grist mills, a sawmill, and eventually a chocolate mill were built there. And, since the mill pond was an attractive facility for baptisms, two Baptist churches were erected near it. A bridge over the mill creek was specified as the only place where butchers could dump their "beasts entrails and garbidg." Mills operated there until about 1804, when they had become less valuable than land that could be created by filling in the pond.

The Hingham, Massachusetts, grain mill, operated on tidal power for nearly three centuries. Tidal waters entered the mill pond, foreground, through the wide tunnel at right, and ran out at the small sluice under the mill, left. *(Hingham Historical Society)*

Tide mills were not yet obsolete, however, and a visionary plan for a huge industrial center was launched in 1813 by a group of promoters, headed by Uriah Cotting. Boston at that time was still a near-island, connected with the mainland by a narrow neck. The tide-washed Charles River made a large indentation on the north side, known as the Great Bay or Back Bay. The promoters proposed to dam the bay, at a cost of $250,000, thereby creating tidal power facilities for eighty-one mills—six grist mills, eight flour mills, six sawmills, sixteen cotton and eighteen woolen mills, twelve rolling and slitting mills, and others to produce cannon, anchors, scythes, grindstones, paints, and such products. "Erect these mills, and lower the price of bread," exhorted the prospectus.

A stone dam fifty feet wide, with a road running over it as an extension of Beacon Street, was built. Another dam perpendicular to it divided a 100-acre Full Basin from a 500-acre Receiving Basin. When the tide was rising, water was admitted to the Full Basin through six gates, which were closed when the tide began to ebb. The water then ran from the Full Basin (until it was empty) through races, in which it turned the mill wheels, into the Receiving Basin. At low tide, water was drained from the Receiving Basin into the Charles through five gates, which closed again as the tide began to rise.

The goal of eighty-one mills was never reached and Boston failed to become a tide-powered Pittsburgh. Two grain mills operated for a time, then were converted to making paints and dyes. An iron foundry and a cordage plant operated. Net receipts for the project in 1834 were only $6133—and $700,000 had been put into construction.

Meanwhile, textile mills using Merrimack River power were thriving, and Lowell and Lawrence became the center of the textile industry. Railroad construction interfered with the operation of the Back Bay mill pond and, more importantly, railways changed the entire transportation system of the country. Steam proved more effective than tidal power and factories were built elsewhere to use it. By 1861 only a drug mill was left at Back Bay. Residents of the area complained about the clouds of fine dust that blew from the Receiving Basin when it was empty.

222

The end result was that the mill pond was filled in and became Boston's finest residential district — still called Back Bay but, in the inevitable evolution of a city, now shadowed by skyscrapers.

There were many other tide mills on the Atlantic coast and they were used for various purposes, from pounding rice in South Carolina and crushing sugar cane in Georgia to sawing logs floated down the rivers of Maine. Peveril Meigs found the sites of more than 300 of them. Most of the mills north of Cape Cod had horizontal tub wheels or turbines, operating below the water's surface and thus avoiding interference from ice in the winter. The traditional vertical wheel was more common in the South.

Some of the mills continued operating into the twentieth century. The Saddle Rock gristmill at Great Neck, Long Island, was restored more than two decades ago and is operated for public viewing by the Nassau County park department. Perhaps the last — or at least nearly the last — mill in business was the Hingham Grain Mill, which I referred to in Chapter 12. It was established in 1643 and continued operating until after the death of its most recent manager in 1938. By that time farms had ceased to be the principal source of flour in Plymouth County but many people still kept chickens, and the mill primarily produced chicken feed, literally and figuratively.

Engineers have long hunted for a way to achieve large-scale utilization of the tides. In 1737 a French artillery officer, Belidor, devised a scheme of using double basins so that impounded water could produce power during the low tide period. In 1849 a proposal was made for a dam and power installation in England's Severn River, where the average tidal range is 28.8 feet. In recent years the Severn (which, it is estimated, could produce 20 trillion watt-hours of electricity a year) has been given renewed study and Solway Firth has also been considered as a power plant site, but Britain has not yet undertaken a project.

Power plants to use the energy of the very high tides in the Bay of Fundy and Passamaquoddy Bay were first proposed nearly sixty years ago and have been considered, off and on, ever since. Through the 1920's an American engineer, Dexter P. Cooper, toiled to get a Passamaquoddy project started. He surveyed 400

The Saddle Brook Grist Mill, Great Neck, Long Island, a tidal mill first mentioned as operating in 1702. *(Nassau County Museum)*

miles of the shore of the bay, which lies at the top of Maine, its waters being divided by the Canadian boundary.

With an area of 100 square miles, mostly in New Brunswick, Passamaquoddy Bay has rugged, irregular shores and its entrance is nearly blocked in some places by rocky islands. Cooper's idea was to build dams between some of the islands, enclosing Passamaquoddy Bay to form an upper basin. This would require Canadian cooperation. The adjacent Cobscook Bay, which has thirty-nine square miles and is entirely in Maine, would be enclosed to form a low pool. Turbines to produce power would be located in races running from the upper to the lower pool. Cooper secured the backing of American power companies, and the future of the project appeared promising until the stock market crash of 1929.

Cooper had a house on Campobello Island, in Canadian waters at the mouth of Passamaquoddy Bay, and it was not far from the summer home of Franklin D. Roosevelt. He won Roosevelt's interest in the project in the twenties; in 1933, after becoming President, F.D.R. had the U.S. Army Corps of Engineers make a study of tidal power. They found it feasible. By 1934 a combination of circumstances made initiation of the project desirable: the Public Works Administration had been formed to create construction jobs; the Eastport, Maine area was hard hit by the Depression and needed help; and jobs might woo staunchly Republican Maine for the Democrats.

Secretary of the Interior Harold L. Ickes went to Maine in August 1934 to look into the project. He went at the request of Roosevelt's secretary, Louis M. Howe, who was concerned about the coming Congressional election. Ickes wrote in his diary that the PWA "does not have the money necessary to finance this project, even if we should approve it, but Louis thought at least a graceful gesture could be made with beneficial results on election day."

By 1935, however, funds were made available by Congress and $10 million (later reduced to $7 million) was alloted for the beginning of construction. Roosevelt asked Cooper, "Can the Army Corps of Engineers handle this project?"

Low tide at Eastport, Maine, where the mean tidal range—18.2 feet—gives an idea of the potential for tidal power generation in the Passamaquoddy Bay region. *(Army Corps of Engineers)*

"Any fool could," Cooper replied. "The hard work has already been done."

So the Engineers made a start on July 5, 1935. In the following January, Ickes wrote:

> Late this afternoon the President called me and he frankly told me that he was in a good deal of a jam. The estimated cost for building the Passamaquoddy project was originally $35 million, and it was on this basis that the President decided to go ahead with it. Recently the Army engineers brought in a supplemental report showing a final cost of about $65 million. The President said that, of course, that made it an entirely different proposition. He sent for George Dern [Secretary of War] and Dern said that he didn't know anything about it, but the estimate of $35 million was accepted as the estimate of Dexter Cooper, the engineer who pioneered the project, and had not been checked until recently. The President told George Dern that he ought to fire all of his Army engineers.

The Corps of Engineers had not been hesitant about preparing for a long-term project. They built Passamaquoddy Village on Moose Island, with 129 housing units for 2400 people and 110 other buildings. (This alienated Eastport villagers, who had hoped to rent quarters to them.)

In March of 1936 Roosevelt admitted that he had a bear by the tail and wanted to let go of it. "I told him that there was a good deal of local dissatisfaction with the way the project was being run," wrote Ickes.

> The Army engineers have bought a $17,000 yacht, and, according to what Dexter Cooper says, they have built seven houses costing $30,000 each. These are in addition to many less expensive houses. The natives of Eastport and that part of Maine seem to think that the

Army engineers are spending money with a lav-
ish hand and, of course, that goes against
Yankee grain.

Meanwhile, with Maine power companies opposed to a TVA-
type public power project, the Maine legislature had failed to
establish a Passamaquoddy Authority to administer the project.
Roosevelt withdrew $5 million in funds. In Congress, an appro-
priation of $9 million to continue the work was tied in to a bill to
provide funds for construction of a ship canal across Florida. The
Maine senators opposed the ship canal, with the result that the
bill was defeated. Work in Maine stopped on June 30, 1936, and
Passamaquoddy Village was turned over to the National Youth
Administration for a training school. During World War II it was
a training camp for Navy Seabees.

Any thoughts Roosevelt had of winning votes in Maine disappeared in the 1936 election. He carried every state except Maine
and Vermont.

From time to time through succeeding years, the Passamaquoddy project raised its head again. In 1948 an International
Joint Commission was directed by the U.S. and Canadian governments to review previous studies and make recommendations for
a new one. The result was a survey, completed in 1959, by an International Passamaquoddy Engineering Board. It envisioned a
powerhouse at Carryingplace Cove with thirty generating units,
which would produce an average of 1843 million kilowatt-hours
a year. Since the peak of energy produced would vary with the
lunar cycle, which is often out of phase with solar-scheduled con-
sumer demand, the report proposed using excess energy to pump
water into a 240,000-acre-foot reservoir in New Brunswick. This
would smooth out the production of power but reduce the annual
output to 1759 million kilowatt-hours.

Another possibility considered was the construction of a
hydroelectric plant on the upper St. John River in Maine, which,
with Passamaquoddy tidal output, would bring the total production of power to 3063 million kilowatt-hours. Market surveys
showed that by 1980 the energy demands of Maine and New

Brunswick would be great enough to absorb the power. The scheme would save about 5,700,000 barrels of oil a year. The total cost would be about $630 million (almost ten times the amount that scared Roosevelt). On a seventy-five-year amortization basis, the cost of power would be 8.4 mills per kilowatt-hour if financed in the U.S. or 11.5 mills if financed in Canada (where the interest rate would be higher). The project would be justified for the U.S. but not for Canada, the report concluded.

In 1961, President Kennedy asked the Department of Interior to review the findings of the international commission. In 1965 Secretary Stewart Udall reported new estimated costs: a total of $900 million for the tidal power project, a two-dam hydroelectric installation on the St. John, and a transmission system.

At these costs, and with rising interest rates, the St. John hydroelectric project would have a benefit–cost ratio of 1.81 and Passamaquoddy only .86 — less than the break-even point of 1. The Secretary recommended that the government go ahead with the St. John project but authorize "continued study, re-examination, and possible redesign" of the Quoddy plan. By this time, nuclear power was coming into its own and proving capable of producing electricity more cheaply.

But since the mid-sixties, nuclear power and the St. John project have run into environmental objections. The St. John project was held up for a time when a discovery was made that damming the river would destroy a unique stand of a rare wildflower, the Furbish lousewort. The Army Corps of Engineers later found the lousewort growing elsewhere but other obstacles remained.

Quoddy is not dead yet. Costs of construction continued to go up but so did the price of oil, with alarming reminders that the supply is dwindling and that the flow from the Middle East can be interrupted by unpredictable circumstances.

In 1977, the Energy Research and Development Administration received a report on a new study which it commissioned Stone & Webster Engineering Corporation of Boston to undertake — a study not only of the tidal potential at Passamaquoddy but also in Cook Inlet in Alaska. By this time, the factors to be

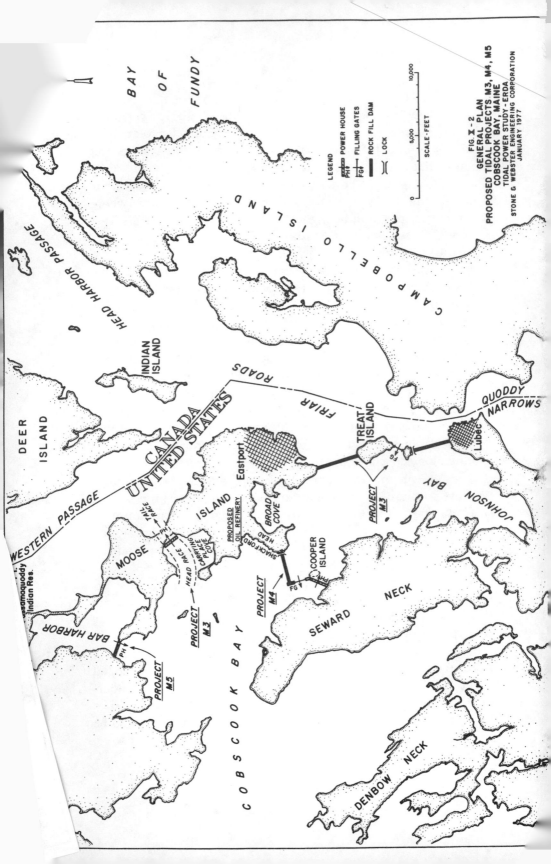

FIG. X-2
GENERAL PLAN
PROPOSED TIDAL PROJECTS M3, M4, M5
COBSCOOK BAY, MAINE
TIDAL POWER STUDY-ERDA
STONE & WEBSTER ENGINEERING CORPORATION
JANUARY 1977

considered had become vastly more complex. What would be the impact on the environment — on waterfowl, on marine life, on historic and archaeological sites? What would be the economic and social impact on the local communities — especially the Passamaquoddy Indians who had gone into litigation to claim a large area in Maine?

At Cook Inlet there were special problems to be taken into account: earthquakes, tsunami, glaciers, landslides, and winter icing. Although it offered a greater tidal potential (with a mean range of 26.1 feet in Knik Arm and 30.3 feet in Turnagain Arm compared to 18.2 feet at Eastport, Maine), various factors — plus doubt that Alaska could use all of the power — led to a conclusion that Passamaquoddy would be a preferred location for a major tidal plant.

Because Canada was contemplating a tidal power project of her own in the Bay of Fundy, the Stone & Webster report recommended that the U.S. concentrate on developing a project in Cobscook Bay, wholly within its own waters. To consider further the use of all Passamaquoddy Bay, in Canadian waters, would cause delay while reaching an international agreement.

This meant, as the report saw it, that the U.S. might have to settle for a single-pool project. Of five possible schemes considered, the most favored one would require dams, with a total length of more than a mile, from Lubec to Treat Island to Eastport — similar to the plan contemplated in 1935. This would impound the waters of Cobscook Bay, which has an area of about thirty-nine square miles (twenty-six square miles at low neap tide). Tidal waters would enter the bay through twenty-five gates in the dams and would exit through a race cut through the narrow isthmus in Moose Island, back of Eastport, where power

(Opposite): This map shows the plan for tidal power generation in the Passamaquoddy Bay area which was favored by the Stone & Webster report of 1977. Cobscook Bay would serve as the single basin, tidewater being impounded by dams from Eastport to Treat Island to Lubec, Maine. The power plant would be situated on a race cut through the narrow part of Moose Island. *(Stone & Webster Engineering Corp.)*

231

would be generated. Generation preferably would occur on the ebbing tides, during two six-hour periods each day when the sea outside would be dropping below the level of the bay. Pumped storage of water would provide a dependable peak capacity. The simplest of several types of plants considered would have a capacity of 180 million watts and a net annual output of 583 million kilowatt-hours. Although the cost of power, 122 mills per kilowatt-hour, would be higher than that for oil- or coal-fired power, the projected continuing increases in the cost of fuel would mean that tidal power would reach the break-even point in thirteen years; over a period of fifty years, the use of tidal power would save 48 million barrels of oil or 18 million tons of coal. Stone & Webster recommended further study.

Further study? Yes, but first the Army Corps of Engineers was asked to conduct a study on how such a study should be made — to produce a "plan of study." This preliminary work was done during 1978 and there was a series of workshops and public hearings to gather the ideas and objections of everyone who might be concerned.

As a basis for discussion, the Engineers suggested three single-pool plans:

A 62.5 million-watt plant with a 30 million-watt auxiliary plant, which together would cost $281.7 million (at 1976 prices).

A 125 million-watt plant to cost $371.8 million.

A 250 million-watt plant to cost $635 million.

Annual operating costs would be about $24 million for the first two schemes and $42 million for the third. "It should be noted," said a prospectus issued by the Engineers,

> that in none of these estimates did the calculation of annual benefits from power sales exceed annual costs.
>
> Compared with fossil-fired plants, which have lower investment costs, tidal power plants should prove considerably more economical on a long-term basis because they are not depen-

dent upon fuel which escalates in price over the
life cycle of the project.

But construction costs continue to rise and no one can be sure how much they will increase during the next decade. One informed guess is that by the time a Cobscook Bay plant can be operating the total cost will be $1.5 billion.

"If everything goes well—if our study is completed in good time and if Congress appropriates funds without delay—1992 is the earliest we could expect the completion of the project, and 1995 would be a more likely date," said an official in the Boston District Office of the Corps of Engineers.

At least American planners can learn something from practical experience with tidal power in France and Russia.

The first large-scale tidal power plant in the world was built by France in the estuary of La Rance, not far upstream from St. Malo. It was undertaken as a pilot project for an ambitious scheme to dam an enormous area outside the nearby Bay of Mont-Saint-Michel. Tides in this area are very large—more than forty-four feet at equinoctial spring tides in La Rance, with a mean range of nearly twenty-eight feet.

Everyone is familiar with pictures of the vast flats at low tide around the famous Mont-Saint-Michel. It was proposed that this area be enclosed with nearly sixty miles of dams running to the Isles of Chausey, where a 12,000 million-watt power plant would be built. The plant would have an annual output of 25 billion kilowatt-hours.

The decision to build the *usine maremotrice* (tidal power station) on La Rance, looking ahead to future construction of the Chausey project, was made at the time of the Suez crisis, when there was uncertainty about the future supply of oil for France and when nuclear power had yet to prove itself. The plant was completed in 1966 and reached full power the following year.

The Rance estuary is more than ten miles long and, at the point selected for the dam, nearly a half-mile wide. The dam had to be eighty-eight feet high, from the bottom of the deepest channel, to impound trillions of cubic feet of water at the highest spring tide.

The tidal power plant at La Rance, France, which has been in operation since 1966, demonstrating the feasibility of such projects. *(Photo by Michel Brigaud for Département d'Electricité de France)*

234

In sluices tunnelling through the dam, twenty-four bulb turbines were installed. These are streamlined units, looking somewhat like jet engines on an airplane but with four-bladed propellers seventeen and a half feet in diameter. The blades are of variable pitch, adjustable for the best efficiency. The bulbs are so big that a man can descend into each one to service it.

This power installation was designed for maximum flexibility. The bulbs can generate electricity either on the ebb or the flow. They can be used as pumps, either to help fill the basin or to lower its level when the incoming tide begins to provide a head outside the dam.

Electricity is transmitted to Paris, Brittany, and the Rennes–Nantes area and the plant can draw power from the last for pumping. A computer is programmed to choose the best mode of operation and output.

As a technical achievement, the Rance plant has been a brilliant success. With a capacity of 240 million watts, it produces more than 580 billion watt-hours of energy a year. Stone & Webster reported that it is not able consistently to produce enough power during peak demand hours, however. During spring tides it generates 2940 million watt-hours per day and during neap tides only 738. The cost of operation in 1975 was 18.3 mills per kilowatt hour, slightly more than that of hydroelectric plants.

The cost of the Rance project was about $100 million. The price of the Chausey project has been estimated at $18 billion — and costs have increased a great deal since that estimate was made. Studies for the project are being continued but the latest thinking is that it would not be economically acceptable unless the cost were less than half of the estimated amount.

The USSR completed a small, 400-kilowatt tidal plant in 1968 at Kislaya Guba, thirty-seven miles north of Murmansk, with a single reversible-flow bulb turbine which is claimed to have an efficiency of 91 percent compared to 80 percent for those at La Rance. The maximum tide at the site is less than twelve feet but this was sufficient, the Russians thought, for such an experimental plant. A particularly interesting feature was that the plant was constructed near Murmansk and floated into place. This technique has been suggested for the Maine project since it avoids the expensive construction of cofferdams. The Soviets have been considering construction of a 5000-million-watt plant on the Gulf of Mezen, near the entrance to the White Sea, as well as two plants with a total capacity of 45,000 million watts on the Sea of Okhotsk at the eastern end of Siberia.

The Republic of Korea is planning the construction of a 400-million-watt tidal plant, probably in Inchon Bay, with completion expected in 1986. The estimated cost is $400 million.

China, according to last (and outdated) reports has built forty small tidal plants and was constructing eighty-eight more. The largest of these, with a 7000-million-watt capacity, was planned for the Tsientang Chiang, where the famous bore (described in Chapter 7) occurs.

Tidal power plants have been considered by Argentina, Australia, Brazil, India, and other countries. The number of possible sites in the world is limited, however.

Canada is blessed with the Bay of Fundy, which has not only extremely high tides but also many rocky-shored, narrow-necked inlets suitable for building dams. When, after years of consideration, serious attention was turned to the possibilities of tidal power more than a decade ago, engineers identified thirty potential sites — with alternative arrangements for some of the sites. Interestingly, four of the sites were in the Annapolis Basin, where the first New World mill operated about 370 years ago.

In a report on the feasibility of tidal power, the number of sites was cut to three by the Atlantic Tidal Power Programming Board, which had been set up to make a study by the federal government of Canada, and by the provincial governments of New Brunswick, and Nova Scotia. All three were in the horn-like upper reaches of the Bay of Fundy — Cobequid Bay, Shepody Bay, and Cumberland Basin.

As fuel costs rose and tidal power became more and more attractive, a Bay of Fundy Tidal Power Review Board was directed to make further studies and to provide "a firm estimate of the cost of tidal energy in relation to its alternative, on which to base a decision to proceed further with detailed investigations and engineering design."

In early 1978 the board's report, titled, "Reassessment of Fundy Tidal Power," was placed before the House of Commons and the legislatures of New Brunswick and Nova Scotia. The principal conclusion: a tidal power project is feasible and design should be started but the cost will be at least $3 billion and some financial problems will have to be solved.

The site recommended by the board is in Cumberland Basin, which is at the border of Nova Scotia and New Brunswick, and would give the two provinces equality in benefits. It has the smallest power potential of the three sites, less than a third of that for Cobequid Bay, which would cost three times as much and take eleven years to build. A Cumberland plant can be completed in seven years and, it is estimated, be functioning by 1990. Shepody Bay would provide 50 percent greater capacity but would cost nearly twice as much and was rejected as economically unfeasible.

Tides in Cumberland Basin have a mean range of 34.4 feet and a spring range of 44.3 feet. A mile-long dam across the basin can impound a tremendous amount of water in the ten-mile basin.

Although there are many schemes for trying to overcome the handicap of tidal peaks and "dead" periods each day, the Canadian board reported "The present studies have confirmed that under economic conditions expected to prevail up to and beyond the end of this century, and the likely expansion programs of the power systems of the Maritime provinces only single basins would attain feasibility." Cumberland might have a "single-effect" operation, generating power probably only on the ebb tide when impounded waters are higher than those outside the dam. It might, however, use the Rance plan of generating both on the ebb and the flow, with turbines that also could be used to pump water with excess electricity.

In any case, a plant with a capacity of 1085 million watts and an annual output of 3423 billion watt-hours was proposed. This would be more than four times as powerful as La Rance, about twice as big as the international Passamaquoddy project considered twenty years ago (minus hydroelectric augmentation), and about five times as big as that now contemplated by the U.S. for Cobscook Bay. It would save 3 million barrels of oil a year and 380,000 tons of coal.

To cut construction costs, the Cumberland powerhouse and sluiceway elements might be constructed elsewhere and floated to the site, as was done at Kislaya Guba, avoiding the expense of cofferdams. This would require careful preparation of the bottom, involving removal of a thick bed of silt and clay from the 100-foot-deep channel.

More formidable than technical problems, which lie well within the competence of the state of the art, are the financial ones. The principal market for power would be the Maritime Integrated System, made up of utilities in Nova Scotia, New Brunswick, and Prince Edward Island. That market is expected to grow sevenfold by the year 2010. Until Maritime could use the full output, it is expected that power would be sold in the United States, as far away as New York City. This market is now five or six times greater than the Maritime market.

Rising costs could be expected to push the price of tidal power to 55 mills per kilowatt-hour by 1990, compared to 48 mills for electricity without tidal power; but by the year 2010 the rate would be 2 mills lower than if there were no tidal power. The benefit–cost ratio would be 1:2 and the break-even period would be thirty to thirty-five years, with seventy-five years regarded as the useful life of the project. By this time, the technology might be obsolete or repairs so costly as to be economically unfeasible.

The Board thus found the Cumberland project financially feasible but concluded that

> year-to-year development costs would make it very difficult, if not impossible, for a utility or a group of utilities in the Maritime provinces to justify a commitment to a tidal power development. . . .
>
> Because the minimum investment would be about $3 billion and because this would result in an inordinate financial burden being placed on

> utility customers in the early years, financial
> feasibility of a tidal power plant would be condi-
> tional upon substantial direct participation by
> governments which would enable the raising of
> the necessary capital and maintaining the cost
> of service to utility customers at annual levels
> not exceeding those which would be incurred by
> an optimal generation expansion program with-
> out tidal power.

The government would have to help foot the bill, and $33 million would be needed for "pre-investment investigations and designs."

So the good news for Canadian citizens was that the Bay of Fundy could produce plenty of power, and the bad news was that tidal power is not free.

Aside from this, there was some good news for New Englanders living along the shore. Construction of a dam at one of the sites in Minas Basin would have raised the height of tides all along the coast — at Boston by one foot. This site was not seriously considered, for it would have required a very large dam. But the Copequid Bay site, which was considered seriously, would have increased tides at Boston by five inches. It may seem strange that a dam nearly 400 miles away would have such a marked effect; but that's what computer analysis showed. As the tides roll over the continental shelf of the Gulf of Maine they carry an enormous amount of energy; if tides are blocked at the head of the Bay of Fundy the energy will make itself felt elsewhere. And what New Englanders, after their experience with coastal flooding, don't need is higher tides.

But it turns out that a dam at the preferred site, Cumberland Basin, will have little, if any effect on the tides at Boston. As for the distant future, if much larger tidal power plants are built, who knows? Perhaps the tides are not immutable after all.

NOTES

Many of the books I have already listed have sections on tidal power but for the most recent developments one must read *Reassessment of Fundy Tidal Power*, the report of the Bay of Fundy Tidal Power Review Board (dated November 1977 but released in March 1978) and *Tidal Power Study for the*

United States Energy Research and Development Administration, Final Report, in two volumes by W. W. Wayne, Jr., for Stone & Webster Engineering Corporation (National Technical Information Service, Department of Commerce, Springfield, Va.). "Fundy Tidal Power," a paper presented in 1977 by R. P. DeLory, manager of the Projects Division, Nova Scotia Power Corporation, was useful. Mr. DeLory and G. C. Baker, executive vice president of the Tidal Power Corporation in Nova Scotia, were helpful in correspondence as well. James E. Callahan, project manager of the tidal power study for the Army Corps of Engineers, New England District, provided materials and sage counsel. Also, Senator Edward M. Kennedy and Rudolph A. Black of the Department of Energy supplied valuable information.

For a survey of tidal power, an indispensable resource is *Tidal Power,* Proceedings of an International Conference on the Utilization of Tidal Power, May 24-29, 1970, at the Atlantic Industrial Research Institute, Nova Scotia Technical College, Halifax, edited by T. J. Gray and O. K. Gashus (Plenum Press, New York, 1972). It contains papers by F. L. Lawton, L. B. Bernshtein, and others. Also useful are: *Investigation of the International Passamaquoddy Tidal Power Project* (Report to the International Joint Commission by the International Passamaquoddy Engineering Board, October 1959); *The International Passamaquoddy Tidal Power Project and Upper Saint John River Hydroelectric Power Development* (Department of the Interior, 1964); Letter to the President from Secretary of the Interior Stewart Udall, July 9, 1965; and *Passamaquoddy Tidal Power Project* by C. Frank Keyser (The Library of Congress, Legislative Reference Service, September 22, 1947). The Library of Congress issued a bibliography, *Tidal Energy* by Jane Collins, in July 1977. *Time & Tide* by F. L. Lawton (*Oceanus*, No. 5, Summer 1974) is an excellent article. *The Secret Diary of Harold L. Ickes, The First Thousand Days, 1933-1936* (Simon and Schuster, New York, 1953) was a source of quotations.

For information about La Rance and Chausey I profusely thank Eliane Morin of the French Embassy for a number of publications: "Les Énergies Nouvelles" by Maurice Magnien (*Defense Nationale*, August–September 1977); "La Rance: Seven Years of Operating a Tidal Powerplant in France" by Johannes Cotillon (*Water Power*, October 1974); "Operating Experience With Bulb Units at the Rance Tidal Power Plant and Other French Hydro-Power Sites" by H. André, Electricité de France; "Les Realisations d'Electricité de France Concernant l'Energie Maremotrice" by R. Bonnefille, Département Echanges Atmosphériques et Pollution; "L'Usine Maremotrice de la Rance," Electricité de France, and a special issue of *Revue Francaise de l'Energie* (No. 183, September–October 1966, Paris) with articles in both French and English. Kim Se-Jong, chief of Power Generating Division, Bureau of Electricity, Republic of Korea, supplied information about Korean tidal power plans. The U.S.S.R. Embassy did not respond to a request for information.

Peveril Meigs of Wayland, Massachusetts, who has a book in preparation on tidal mills, on which he is the leading authority, was most generous in his assistance. His publications include "Energy In Early Boston" (*New England Historical and Genealogical Register*, Vol. CXXVIII, April 1974) and "Tide Mills on the North Atlantic Coast" (*Annals of the Association of American Geographers*, Vol. 59, March 1969). *Boston, A Topographical History* by Walter Muir Whitehill (Belknap Press of Harvard University Press, Cambridge, 1968) has useful information, as do *Early American Mills* by Martha and Murray Zimiles (Clarkson N. Potter, New York, 1973) and *Windmills & Watermills* by John Reynolds (Praeger, New York, 1970). I thank Volta Torrey, author of *Wind-Catchers* (Stephen Greene Press, Brattleboro, Vermont), James Wheaton and Michael Shilhan of the Hingham Historical Society and Robert Fraser, curator of the Cohasset Historical Society, for their kind help.

Index

243

Knot, ix

Lamar, D. L., 130
Laplace, Pierre Simon, 29, 120
Lee, Albert, 10
Leet, L. Don, 139
Leggett, William C., 181-2
Leonardo da Vinci, 18
Lieber, Arnold L., 199
Lubbock, Sir William, 90
Lyman, Henry, 184

MacArthur, Gen. Douglas, 214
MacDonald, Gordon J. F., 131, 150
McGowan, Brian, 73
Macmillan, D. H., 204-205
Maelstroms, 113-14
Mangroves, 173-5
Manilius, 5
Man-of-Signs. *See* Zodiac
Marine Biological Laboratory, 175, 190, 198
Mark, Alexandra, 7
Marmer, H. A., 47, 106-107
Marshack, Alexander, 4
Marshes, 171-6
Marvin, Ursula B., 125
Maryland, University of, 24
Massachusetts Institute of Technology (M.I.T.), 25, 54, 116, 150, 155
Massachusetts, University of, 75, 190
Mather, Cotton, 9, 64-5
Meeus, Jean, 148
Meigs, Peveril, 220, 223
Melchior, Paul, 118
Menaker, Abraham, 201
Menaker, Walter, 201
Menkin, Miriam F., 201
Menstruation, 2-3, 200-202
Merifield, P. M., 130
Michelson, A. A., 25
Moon, 124-35 (*ill.* 126)
    apogee, 34
    geology, 132-4
    gravitational force, 8, 29-30
    moonquakes, 134-5
    origin, 128-131
    perigee, 34
    tidal friction, 128-9, 131-2
    and weather, 10
Moon's Man. *See* Zodiac
Moore, Captain, 111
Moore, John Hamilton, 88
More, Louis Trenchard, 20-21

Morgan, Elaine, 202
Morison, Samuel Eliot, 3, 204, 208-209
Moscow Univeristy, 24
Motz, Lloyd, 131-2
Muir, E. A., 186
Munk, Walter, H., 150

National Aeronautics and Space Administration (NASA), 25, 117
National Environmental Satellite Service, 85
National Flood Insurance Program, 74-5
National Geodetic Survey, Office of, 86
National Marine Fisheries Service, 85
National Oceanic and Atmospheric Administration (NOAA), 69, 85-6, 92-3, 99, 164-5
National Ocean Survey (NOS), 57, 60, 85-6, 92-9
National Weather Service, 85, 92-3, 164-5
Nautical Almanac Office, 146
Naval Observatory, 61, 146-8 (*ill.* 147)
Northwestern University, 195
Newton, Isaac, 20-21
Nutations, 149-50

Okal, Emile, 145-6

Palmer, John D., 190-91
Palola worms, 191-2
Paul the Deacon, 17
Perrey, Alexis, 140
Perry, Robert Elton, 89
Petterson, Otto, 54
Plagemann, Stephen H., 142-9
Planetary Science Institute, 127
Pliny the Elder, 1, 17, 118
Plutarch, 18, 124
Poe, Edgar Allan, 113
Pollution, 176-9
Posidonius, 17
Press, Frank, 150, 155
Princeton University, 26
Ptolemy (Claudius Ptolemaeus), 5
Pytheas of Marseille, 17

Qazvini, Zakariyya, 18
Quadrature, 33

Redfield, Alfred C., 47-8
Resonance, 45-6
Reversing Falls, 112
Rochester, University of, 24